CAMBRIDGE LIBRARY COLLECTION

Books of enduring scholarly value

Physical Sciences

From ancient times, humans have tried to understand the workings of
the world around them. The roots of modern physical science go back to
the very earliest mechanical devices such as levers and rollers, the mixing
of paints and dyes, and the importance of the heavenly bodies in early
religious observance and navigation. The physical sciences as we know them
today began to emerge as independent academic subjects during the early
modern period, in the work of Newton and other 'natural philosophers',
and numerous sub-disciplines developed during the centuries that followed.
This part of the Cambridge Library Collection is devoted to landmark
publications in this area which will be of interest to historians of science
concerned with individual scientists, particular discoveries, and advances in
scientific method, or with the establishment and development of scientific
institutions around the world.

Experiments and Observations on Electricity

The political legacy of Benjamin Franklin (1706–90) has tended to
overshadow his inventions and scientific discoveries, but this work, of which
the 1769 enlarged fourth edition is reissued here, gathers together published
and unpublished letters which demonstrate the range of his interests. A large
number (many addressed to his friend, the English botanist Peter Collinson)
are about experiments with electricity, but (as the subtitle, 'to which are
added letters and papers on philosophical subjects', makes clear) others
discuss the setting of words to music, the warming of water by pumping, and
how the speed of a boat varies with the depth of the water it moves through.
The letters (and some replies) show a lively transatlantic group of scientific
friends and colleagues describing their experiments, interpreting each others'
results, and theorizing on all aspects of the natural world. Franklin's three-
volume autobiography is also reissued in the Cambridge Library Collection.

Cambridge University Press has long been a pioneer in the reissuing of out-of-print titles from its own backlist, producing digital reprints of books that are still sought after by scholars and students but could not be reprinted economically using traditional technology. The Cambridge Library Collection extends this activity to a wider range of books which are still of importance to researchers and professionals, either for the source material they contain, or as landmarks in the history of their academic discipline.

Drawing from the world-renowned collections in the Cambridge University Library and other partner libraries, and guided by the advice of experts in each subject area, Cambridge University Press is using state-of-the-art scanning machines in its own Printing House to capture the content of each book selected for inclusion. The files are processed to give a consistently clear, crisp image, and the books finished to the high quality standard for which the Press is recognised around the world. The latest print-on-demand technology ensures that the books will remain available indefinitely, and that orders for single or multiple copies can quickly be supplied.

The Cambridge Library Collection brings back to life books of enduring scholarly value (including out-of-copyright works originally issued by other publishers) across a wide range of disciplines in the humanities and social sciences and in science and technology.

Experiments and Observations on Electricity

Made at Philadelphia in America

BENJAMIN FRANKLIN

CAMBRIDGE
UNIVERSITY PRESS

University Printing House, Cambridge, CB2 8BS, United Kingdom

Cambridge University Press is part of the University of Cambridge.

It furthers the University's mission by disseminating knowledge in the pursuit of
education, learning and research at the highest international levels of excellence.

www.cambridge.org
Information on this title: www.cambridge.org/9781108080163

© in this compilation Cambridge University Press 2019

This edition first published 1769
This digitally printed version 2019

ISBN 978-1-108-08016-3 Paperback

This book reproduces the text of the original edition. The content and language reflect
the beliefs, practices and terminology of their time, and have not been updated.

Cambridge University Press wishes to make clear that the book, unless originally published
by Cambridge, is not being republished by, in association or collaboration with,
or with the endorsement or approval of, the original publisher or its successors in title.

The original edition of this book contains a number of oversize plates
which it has not been possible to reproduce to scale in this edition.
They can be found online at www.cambridge.org/9781108080163

EXPERIMENTS

AND

OBSERVATIONS

ON

ELECTRICITY,

MADE AT

PHILADELPHIA in AMERICA.

EXPERIMENTS

AND

OBSERVATIONS

ON

ELECTRICITY,

MADE AT

PHILADELPHIA in AMERICA,

BY

BENJAMIN FRANKLIN, L. L. D. and F. R. S.

To which are added,

LETTERS and PAPERS

ON

PHILOSOPHICAL SUBJECTS.

The Whole corrected, methodized, improved, and now first collected into one Volume,

AND

Illustrated with COPPER PLATES.

LONDON:

Printed for DAVID HENRY ; and sold by FRANCIS NEWBERY,
at the Corner of St. Paul's Church-Yard.

MDCCLXIX.

ADVERTISEMENT

Concerning this Fourth Edition.

ALL the Philofophical Letters and Papers of the fame Author, that have been inferted at different Times in the Philofophical Tranfactions of the Royal Society, or in the Magazines, or printed in feparate Pamphlets, are collected and added to this Edition ; together with a Number of others on various Subjects, never before printed, that have paffed between the Author and his Friends. Many Errors in the preceding Editions, are now corrected ; fome of the Letters, which had been tranfpofed, are reftored to their proper places ; and fundry Paffages are more fully explained by Notes ----There is alfo added, a compleat Index to the whole.

ERRATA.

PAGE 13, *Line* 8, *for,* at top, *read,* at the top.

50,——15, *add,* becaufe the blood and other humours containing fo much water, are more ready conductors

62——10 *for* wrote, *read* written.

73,——10, *for* prevades, *read,* pervades.

96,—— 7, *from the bottom*: *add,* (page 54 of this Edit.

127,——22, *for,* difcoveries, you *read* difcoveries, you.

134,—— 3, *for Lyden, reat Leyden.*

141,——20, *for* give it a fhock, *read,* give a fhock.

143,——11, *for* experimen, *read,* experiment.

158,——22, *for* they, *read,* you.

189, —— 5, *for* infenfibly, *read,* .infenfibly.

199,——15, *for* fettle, *read* fettle.

225,—— 3, *add,* See plate II.

327—— 7, *for* by frequent, *read,* by the frequent.

344,—— 3, *from the bottom, begin a new paragraph at* There.

345,—— 4, *from the bottom, begin a new paragraph at* How.

346——20, *dele the comma at* reproduced, *and put one at* feparation.

347,——11, *for* has the fame, *read,* has nearly the fame.

354,—— 2, *from the bottom, for* 280, *read* 180.

389——11, *add* See plate VI.

403,——15, *for* VI. *read* VII.

475,—— 2, *for* endeabour, *read* endeavour.

490,——18, *dele the femicolon between* warmer *and* fituation.

In numbering of the pages, 112, 113, are repeated; as are alfo pages 465 to 472.

PREFACE

To the First Edition.

IT may be neceffary to acquaint the Reader, that the following obfervations and experiments were not drawn up with a view to their being made publick, but were communicated at different times, and moft of them in letters wrote on various topicks, as matters only of private amufement.

But fome perfons to whom they were read, and who had themfelves been converfant in electrical difquifitions, were of opinion, they contained fo many curious and interefting particulars relative to this affair, that it would be doing a kind of injuftice to the public, to confine them folely to the limits of a private acquaintance.

The Editor was therefore prevailed upon to commit fuch extracts of letters, and other detached pieces as were in his hands to the prefs, without waiting for the ingenious author's permiffion fo to do; and this was done with the lefs hefitation, as it was apprehended the author's engagements in other affairs would fcarce afford him leifure to give the publick his reflections and experiments on the fubject, finifhed with that care and precifion, of which the treatife before us fhews he is alike ftudious and capable.

The experiments which our author relates are moft of them peculiar to himfelf; they are conducted with judgment, and the inferences from them plain and conclufive; though fometimes propofed under the terms of fuppofitions and conjectures.

And indeed the fcene he opens, ftrikes us with a pleafing aftonifhment, while he conducts us by a train of facts and judicious reflections, to a probable caufe of thofe phænomena, which are at once the moft awful, and, hitherto, accounted for with the leaft verifimilitude.

He

He exhibits to our consideration, an invisible, subtle matter, disseminated through all nature in various proportions, equally unobserved, and, whilst all those bodies to which it peculiarly adheres are alike charged with it, inoffensive.

He shews, however, that if an unequal distribution is by any means brought about; if there is a coacervation in one part of space, a less proportion, vacuity, or want, in another; by the near approach of a body capable of conducting the coacervated part to the emptier space, it becomes perhaps the most formidable and irresistible agent in the universe. Animals are in an instant struck breathless, bodies almost impervious by any force yet known, are perforated, and metals fused by it, in a moment.

From the similar effects of lightning and electricity, our author has been led to make some probable conjectures on the cause of the former; and, at the same time, to propose some rational experiments in order to secure ourselves, and those things on which its force is often directed, from its pernicious effects; a circumstance of no small importance to the publick, and therefore worthy of the utmost attention.

It has, indeed, been of late the fashion to ascribe every grand or unusual operation of nature, such as lightning and earthquakes, to electricity; not, as one would imagine from the manner of reasoning on these occasions, that the authors of these schemes have discovered any connection betwixt the cause and effect, or saw in what manner they were related; but, as it would seem, merely because they were unacquainted with any other agent, of which it could not positively be said the connection was impossible.

But of these, and many other interesting circumstances, the reader will be more satisfactorily informed in the following letters, to which he is therefore referred by

The E D I T O R.

Plate. 1.

Fig. I.   Fig. II.   Fig. III.   Fig. IV.

Fig. V.   Fig. IX.

Fig. X.

Fig. VI.

Fig. VII.

Fig. VIII.

EXTRACT

OF

LETTER I.

FROM

BENJ. FRANKLIN, *Esq*; at *Philadelphia*,

TO

PETER COLLINSON, Esq; F. R. S. *London*.

S I R,   *Philadelphia, March* 28, 1747.

YOUR kind prefent of an electric tube, with directions for ufing it, has put feveral of us on making electrical experiments, in which we have obferved fome particular phænomena that we look upon to be new. I fhall, therefore communicate them to you in my next, though pof-

B   fibly,

fibly they may not be new to you, as among the numbers
daily employed in thofe experiments on your fide the water, 'tis probable fome one or other has hit on the fame obfervations.   For my own part, I never was before engaged
in any ftudy that fo totally engroffed my attention and my
time as this has lately done ; for what with making experiments when I can be alone, and repeating them to my
Friends and Acquaintance, who, from the novelty of the
thing, come continually in crouds to fee them, I have, during fome months paft, had little leifure for any thing elfe.

I am, &c.

B. FRANKLIN.

LET-

L E T T E R II.

F R O M

Mr BENJ. FRANKLIN, in *Philadelphia*,

T O

PETER COLLINSON, Efq; F. R. S. *London.*

S I R,   *July* 11, 1747.

IN my laft I informed you that, in purfuing our electrical enquiries, we had obferved fome particular Phænomena, which we looked upon to be new, and of which I
promifed to give you fome account, though I apprehended
they might poffibly not be new to you, as fo many hands
are daily employed in electrical experiments on your fide
the water, fome or other of which would probably hit on
the fame obfervations.

The firft is the wonderful effect of pointed bodies, both in
drawing off and *throwing off* the electrical fire. For example,

Place an iron fhot of three or four inches diameter on the
mouth of a clean dry glafs bottle. By a fine filken thread
from the cieling, right over the mouth of the bottle, fufpend a fmall cork-ball, about the bignefs of a marble; the

thread of fuch a length, as that the cork-ball may reft
againft the fide of the fhot. Electrify the fhot, and the
ball will be repelled to the diftance of four or five inches,
more or lefs, according to the quantity of Electricity.——
When in this ftate, if you prefent to the fhot the point of
a long flender fharp bodkin, at fix or eight inches diftance,
the repellency is inftantly deftroy'd, and the cork flies to
the fhot. A blunt body muft be brought within an inch,
and draw a fpark, to produce the fame effect. To prove
that the electrical fire is *drawn off* by the point, if you take
the blade of the bodkin out of the wooden handle, and fix
it in a ftick of fealing-wax, and then prefent it at the diftance aforefaid, or if you bring it very near, no fuch effect
follows ; but fliding one finger along the wax till you
touch the blade, and the ball flies to the fhot immediately.
—If you prefent the point in the dark, you will fee, fometimes at a foot diftance, and more, a light gather upon it,
like that of a fire-fly, or glow-worm ; the lefs fharp the
point, the nearer you muft bring it to obferve the light ;
and at whatever diftance you fee the light, you may draw
off the electrical fire, and deftroy the repellency.—If a
cork-ball fo fufpended be repelled by the tube, and a point
be prefented quick to it, tho' at a confiderable diftance,
'tis furprizing to fee how fuddenly it flies back to the tube.
Points of wood will do near as well as thofe of iron, provided the wood is not dry ; for perfectly dry wood will no
more conduct Electricity than fealing-wax.

To

To fhew that points will *throw off* * as well as *draw off* the electrical fire ; lay a long fharp needle upon the fhot, and you cannot electrife the fhot, fo as to make it repel the cork-ball †.—Or fix a needle to the end of a fufpended gun-barrel, or iron-rod, fo as to point beyond it like a little bayonet ; and while it remains there, the gun-barrel, or rod, cannot by applying the tube to the other end be elec-trifed fo as to give a fpark, the fire continually running out filently at the point.　In the dark you may fee it make the fame appearance as it does in the cafe before-mentioned.

The repellency between the cork-ball and the fhot is likewife deftroy'd.　1. By fifting fine fand on it ; this does it gradually.　2. By breathing on it.　3. By making a fmoke about it from burning wood ‡.　4. By candle light, even though the candle is at a foot diftance : thefe do it fuddenly.—The light of a bright coal from a wood fire ; and

* This power of points to *throw off* the electrical fire, was firft com-municated to me by my ingenious friend Mr *Thomas Hopkinfon*, fince deceafed, whofe virtue and integrity, in every ftation of life, public and private, will ever make his Memory dear to thofe who knew him, and knew how to value him.

† This was Mr *Hopkinfon's* Experiment, made with an expectation of drawing a more fharp and powerful fpark from the point, as from a kind of focus, and he was furprized to find little or none.

‡ We fuppofe every particle of fand, moifture, or fmoke, being firft at-tracted and then repelled, carries off with it a portion of the electrical fire ; but that the fame ftill fubfifts in thofe particles, till they communicate it to fomething elfe, and that it is never really deftroyed.——So when water is thrown on common fire, we do not imagine the element is thereby deftroyed or annihilated, but only difperfed, each particle of water carrying off in vapour its portion of the fire, which it had attracted and attached to itfelf.

the

the light of red-hot iron do it likewife ; but not at fo great a diftance. Smoke from dry rofin dropt on hot iron, does not deftroy the repellency ; but is attracted by both·fhot and cork-ball, forming proportionable atmofpheres round them, making them look beautifully, fomewhat like fome of the figures in *Burnet*'s or *Whifton*'s theory of the earth.

N. B. This experiment fhould be made in a clofet, where the air is very ftill, or it will be apt to fail.

The light of the fun thrown ftrongly on both cork and fhot by a looking-glafs for a long time together, does not impair the repellency in the leaft. This difference between fire-light and fun-light is another thing that feems new and extraordinary to us *.

We had for fome time been of opinion, that the electrical fire was not created by friction, but collected, being really an element diffus'd among, and attracted by other matter, particularly by water and metals. We had even difcovered and demonftrated its afflux to the electrical fphere, as well as its efflux, by means of little light windmill wheels made of ftiff paper vanes, fixed obliquely and turning freely on fine wire axes. Alfo by little wheels of the fame matter, but formed like water-wheels. Of the

---

* This different Effect probably did not arife from any difference in the light, but rather from the particles feparated from the candle, being firft attracted and then repelled, carrying off the electric matter with them; and from the rarefying the air, between the glowing coal or red-hot iron, and the electrifed fhot, through which rarified air the electric fluid could more readily pafs.

difpofition and application of which wheels, and the various phænomena refulting, I could, if I had time, fill you a fheet †. The impoffibility of electrifing one's felf (though ftanding on wax) by rubbing the tube, and drawing the fire from it ; and the manner of doing it, by paffing the tube near a perfon or thing ftanding on the floor, &c. had alfo occurred to us fome months before Mr *Watfon*'s ingenious *Sequel* came to hand, and thefe were fome of the new things I intended to have communicated to you.—— But now I need only mention fome particulars not hinted in that piece, with our reafonings thereupon; though perhaps the latter might well enough be fpared.

　　1. A perfon ftanding on wax, and rubbing the tube, and another perfon on wax drawing the fire, they will both of them, (provided they do not ftand fo as to touch one another) appear to be electrifed, to a perfon ftanding on the floor ; that is, he will perceive a fpark on approaching each of them with his knuckle.

　　2. But if the perfons on wax touch one another during the exciting of the tube, neither of them will appear to be electrifed.

　　3. If they touch one another after exciting the tube, and drawing the fire as aforefaid, there will be a ftronger

† Thefe experiments with the wheels were made and communicated to me by my worthy and ingenious friend Mr *Philip Syng* ; but we afterwards difcovered that the motion of thofe wheels was not owing to any afflux or efflux of the electric fluid, but to various circumftances of attraction and repulfion.　1750.

ſpark between them, than was between either of them and the perſon on the floor.

4. After ſuch ſtrong ſpark, neither of them diſcover any electricity.

Theſe appearances we attempt to account for thus : We ſuppoſe, as aforeſaid, that electrical fire is a common element, of which every one of the three perſons abovementioned has his equal ſhare, before any operation is begun with the tube. *A*, who ſtands on wax and rubs the tube, collects the electrical fire from himſelf into the glaſs ; and his communication with the common ſtock being cut oft by the wax, his body is not again immediately ſupply'd. *B*, (who ſtands on wax likewiſe) paſſing his knuckle along near the tube, receives the fire which was collected by the glaſs from *A* ; and his communication with the common ſtock being likewiſe cut off, he retains the additional quantity received.—To *C*, ſtanding on the floor, both appear to be electriſed : for he having only the middle quantity of electrical fire, receives a ſpark upon approaching *B*, who has an over quantity ; but gives one to *A*, who has an under quantity. If *A* and *B* approach to touch each other, the ſpark is ſtronger, becauſe the difference between them is greater : After ſuch touch there is no ſpark between either of them and *C*, becauſe the electrical fire in all is reduced to the original equality. If they touch while electriſing, the equality is never deſtroy'd, the fire only circulating. Hence have ariſen ſome new terms among us : we ſay, *B*, (and bodies like circumſtanced) is electriſed

poſitively ;

positively ; *A, negatively*.   Or rather, *B* is electrifed *plus*;
A, minus. And we daily in our experiments electrife
bodies *plus* or *minus*, as we think proper.—To electrife
plus or *minus*, no more needs to be known than this, that
the parts of the tube or sphere that are rubbed, do, in the
instant of the friction, attract the electrical fire, and therefore take it from the thing rubbing : the same parts immediately, as the friction upon them ceases, are disposed to
give the fire they have received, to any body that has less.
Thus you may circulate it, as Mr *Watson* has shewn ; you
may also accumulate or subtract it upon, or from any body,
as you connect that body with the rubber or with the receiver, the communication with the common stock being
cut off.   We think that ingenious gentleman was deceived
when he imagined (in his *Sequel*) that the electrical
fire came down the wire from the cieling to the gun-
barrel, thence to the sphere, and so electrifed the machine
and the man turning the wheel, *&c.*   We suppose it was
driven off, and not brought on through that wire ; and that
the machine and man, *&c.* were electrifed *minus* ; *i. e.* had
less electrical fire in them than things in common.

As the vessel is just upon sailing, I cannot give you so
large an account of *American* Electricity as I intended : I
shall only mention a few particulars more.—We find gra-
nulated lead better to fill the phial with, than water, being
easily warmed, and keeping warm and dry in damp air.—
We fire spirits with the wire of the phial.——We light
candles, just blown out, by drawing a spark among the

C   smoke

smoke between the wire and snuffers.——We represent lightning, by passing the wire in the dark, over a china plate that has gilt flowers, or applying it to gilt frames of looking-glasses, &c. We electrise a person twenty or more times running, with a touch of the finger on the wire, thus: He stands on wax. Give him the electrised bottle in his hand. Touch the wire with your finger, and then touch his hand or face; there are sparks every time *.——We increase the force of the electrical kiss vastly, thus: Let *A* and *B* stand on wax; or *A* on wax, and *B* on the floor; give one of them the electrised phial in hand; let the other take hold of the wire; there will be a small spark; but when their lips approach, they will be struck and shock'd. The same if another gentleman and lady, *C* and *D*, standing also on wax, and joining hands with *A* and *B*, salute or shake hands. We suspend by fine silk thread a counterfeit spider, made of a small piece of burnt cork, with legs of linnen thread, and a grain or two of lead stuck in him, to give him more weight. Upon the table over which he hangs, we stick a wire upright, as high as the phial and wire, two or three inches from the spider: then we animate him, by setting the electrified phial at the same distance on the other side of him; he will immediately fly to the wire of the phial, bend his legs

---

* By taking a spark from the wire, the electricity within the bottle is diminished; the outside of the bottle then draws some from the person holding it, and leaves him in the negative state. Then when his hand or face is touch'd, an equal quantity is restored to him from the person touching.

in touching it; then fpring off, and fly to the wire in the
table; thence again to the wire of the phial, playing with
his legs againft both, in a very entertaining manner, appearing perfectly alive to perfons unacquainted. He will
continue this motion an hour or more in dry weather.—We
electrify, upon wax in the dark, a book that has a double
line of gold round upon the covers, and then apply a
knuckle to the gilding; the fire appears every where upon
the gold like a flafh of lightning: not upon the leather,
nor, if you touch the leather inftead of the gold. We rub
our tubes with buckfkin, and obferve always to keep the
fame fide to the tube, and never to fully the tube by handling; thus they work readily and eafily, without the leaft
fatigue, efpecially if kept in tight pafteboard cafes, lined
with flannel, and fitting clofe to the tube *. This I mention becaufe the *European* papers on Electricity, frequently
fpeak of rubbing the tube, as a fatiguing exercife. Our
fpheres are fixed on iron axes, which pafs through them.
At one end of the axis there is a fmall handle, with which
you turn the fphere like a common grindftone. This we
find very commodious, as the machine takes up but little
room, is portable, and may be enclofed in a tight box, when
not in ufe. Tis true, the fphere does not turn fo fwift as
when the great wheel is ufed: but fwiftnefs we think of little
importance, fince a few turns will charge the phial, &c.
fufficiently †.   *I am, &c.*   B. FRANKLIN.

* Our tubes are made here of green glafs, 27 or 30 inches long, as big
as can be grafped,
† This fimple eafily-made machine was a contrivance of Mr *Syng*'s.

LETTER III.

FROM

BENJ. FRANKLIN, *Esq*; at *Philadelphia,*

TO

PETER COLLINSON, F.R.S. *London.*

SIR, *Sept.* 1, 1747.

THE neceffary trouble of copying long letters, which, perhaps, when they come to your hands, may contain nothing new, or worth your reading, (fo quick is the progrefs made with you in Electricity) half difcourages me from writing any more on that fubject. Yet I cannot forbear adding a few obfervations on M. *Mufchenbroek*'s wonderful bottle.

1. The non-electric contain'd in the bottle differs when electrifed from a non-electric electrifed out of the bottle, in this: that the electrical fire of the latter is accumulated *on its furface,* and forms an electrical atmofphere round it of

con-

conſiderable extent ; but the electrical fire is crowded *into the ſubſtance* of the former, the glaſs confining it *.

2. At the ſame time that the wire and top of the bottle, &c. is electriſed *poſitively* or *plus,* the bottom of the bottle is electriſed *negatively* or *minus,* in exact proportion : *i. e.* whatever quantity of electrical fire is thrown in at top, an equal quantity goes out of the bottom †.   To underſtand this, ſuppoſe the common quantity of electricity in each part of the bottle, before the operation begins, is equal to 20 ; and at every ſtroke of the tube, ſuppoſe a quantity equal to 1 is thrown in; then, after the firſt ſtroke, the quantity contain'd in the wire and upper part of the bottle will be 21, in the bottom 19.   After the ſecond, the upper part will have 22, the lower 18, and ſo on, till, after 20 ſtrokes, the upper part will have a quantity of electrical fire equal to 40, the lower part none : and then the operation ends : for no more can be thrown into the upper part, when no more can be driven out of the lower part.   If you attempt to throw more in, it is ſpued back through the wire, or flies out in loud cracks through the ſides of the bottle.

3. The equilibrium cannot be reſtored in the bottle by *inward* communication or contact of the parts ; but it muſt be done by a communication form'd *without* the

---

* See this opinion rectified in Letter IV. § 16 and 17.   The fire in the bottle was found by ſubſequent experiments not to be contained in the non-electric, but *in the glaſs.*   1748.

† What is ſaid here, and after, of the *top* and *bottom* of the bottle, is true of the *inſide* and *outſide* ſurfaces, and ſhould have been ſo expreſſed.

bottle

bottle between the top and bottom, by some non-electric, touching or approaching both at the same time ; in which case it is restored with a violence and quickness inexpressible ; or, touching each alternately, in which case the equilibrium is restored by degrees.

4. As no more electrical fire can be thrown into the top of the bottle, when all is driven out of the bottom, so in a bottle not yet electrified, none can be thrown into the top, when none *can* get out at the bottom ; which happens either when the bottom is too thick, or when the bottle is placed on an electric *per se*. Again, when the bottle is electrified, but little of the electrical fire can be *drawn out* from the top, by touching the wire, unless an equal quantity can at the same time *get in* at the bottom *. Thus, place an electrified bottle on clean glass or dry wax, and you will not, by touching the wire, get out the fire from the top. Place it on a non-electric, and touch the wire, you will get it out in a short time ; but soonest when you form a direct communication as above.

So wonderfully are these two states of Electricity, the *plus* and *minus*, combined and balanced in this miraculous bottle ! situated and related to each other in a manner that I can by no means comprehend ! If it were possible that a bottle should in one part contain a quantity of air strongly comprest, and in another part a perfect vacuum, we know the equilibrium would be instantly restored *within*. But

* See the preceding note, relating to *top* and *bottom*.

here

here we have a bottle containing at the fame time a *plenum* of electrical fire, and a *vacuum* of the fame fire; and yet the equilibrium cannot be reftored between them but by a communication *without !* though the *plenum* preffes violently to expand, and the hungry vacuum feems to attract as violently in order to be filled.

5. The fhock to the nerves (or convulfion rather) is occafioned by the fudden paffing of the fire through the body in its way from the top to the bottom of the bottle. The fire takes the fhorteft courfe, as Mr *Watfon* juftly obferves : But it does not appear from experiment that in order for a perfon to be fhocked, a communication with the floor is neceffary : for he that holds the bottle with one hand, and touches the wire with the other, will be fhock'd as much, though his fhoes be dry, or even ftanding on wax, as otherwife. And on the touch of the wire (or of the gun-barrel, which is the fame thing) the fire does not proceed from the touching finger to the wire, as is fuppofed, but from the wire to the finger, and paffes through the body to the other hand, and fo into the bottom of the bottle.

EXPERIMENTS *confirming the above.*

EXPERIMENT I.

Place an electrifed phial on wax; a fmall cork-ball fufpended by a dry filk-thread held in your hand, and
brought

brought near to the wire, will firſt be attracted, and then repelled : when in this ſtate of repellency, ſink your hand, that the ball may be brought towards the bottom of the bottle ; it will be there inſtantly and ſtrongly attracted, 'till it has parted with its fire.

If the bottle had a *poſitive* electrical atmoſphere, as well as the wire, an electrified cork would be repelled from one as well as from the other.

EXPERIMENT II.

FIG. 1. From a bent wire (*a*) ſticking in the table, let a ſmall linen thread (*b*) hang down within half an inch of the electriſed phial (*c*). Touch the wire of the phial repeatedly with your finger, and at every touch you will ſee the thread inſtantly attracted by the bottle. (This is beſt done by a vinegar cruet, or ſome ſuch belly'd bottle.) As ſoon as you draw any fire out from the upper part, by touching the wire, the lower part of the bottle draws an equal quantity in by the thread.

EXPERIMENT III.

FIG. 2. Fix a wire in the lead, with which the bottom of the bottle is armed (*d*) ſo as that bending upwards, its ring end may be level with the top or ring-end of the wire in the cork (*e*), and at three or four inches diſtance. Then electriciſe the bottle, and place it on wax. If a cork ſuſpended by a ſilk thread (*f*) hang between theſe two wires, it will play inceſſantly from one to the other, 'till the bottle

is

is no longer electrifed ; that is, it fetches and carries fire
from the top to the bottom * of the bottle, 'till the equilibrium is reftored.

EXPERIMENT IV.

Fig. 3. Place an electrifed phial on wax; take a wire
(*g*) in form of a *C*, the ends at fuch a diftance when bent,
as that the upper may touch the wire of the bottle, when
the lower touches the bottom : ftick the outer part on a
ftick of fealing-wax (*h*), which will ferve as a handle;
then apply the lower end to the bottom of the bottle,
and gradually bring the upper end near the wire in the
cork. The confequence is, fpark follows fpark till the
equilibrium is reftored. Touch the top firft, and on approaching the bottom with the other end, you have a conftant ftream of fire from the wire entering the bottle.
Touch the top and bottom together, and the equilibrium
will inftantly be reftored ; the crooked wire forming the
communication.

EXPERIMENT V.

Fig. 4. Let a ring of thin lead, or paper, furround a
bottle (*i*) even at fome diftance from or above the bottom.
From that ring let a wire proceed up, till it touch the
wire of the cork (*k*). A bottle fo fixt cannot by any means
be electrifed : the equilibrium is never deftroyed : for

* *i. e.* from the infide to the outfide.

D

while the communication between the upper and lower parts of the bottle is continued by the outfide wire, the fire only circulates : what is driven out at bottom, is conftantly fupply'd from the top †.   Hence a bottle cannot be electrifed that is foul or moift on the outfide, if fuch moifture continue up to the cork or wire.

E X P E R I M E N T   VI.

Place a man on a cake of wax, and prefent him the wire of the electrified phial to touch, you ftanding on the floor, and holding it in your hand.   As often as he touches it, he will be electrified *plus* ; and any one ftanding on the floor may draw a fpark from him.   The fire in this experiment paffes out of the wire into him ; and at the fame time out of your hand into the bottom of the bottle.

E X P E R I M E N T   VII.

Give him the electrical phial to hold ; and do you touch the wire ; as often as you touch it he will be electrified *minus*, and may draw a fpark from any one ftanding on the floor.   The fire now paffes from the wire to you, and from him into the bottom of the bottle.

E X P E R I M E N T   VIII.

Lay two books on two glaffes, back towards back, two or three inches diftant.   Set the electrified phial on one, and then touch the wire ; that book will be electrified

* See the preceding note.

minus

minus ; the electrical fire being drawn out of it by the bottom of the bottle. Take off the bottle, and holding it in your hand, touch the other with the wire ; that book will be electriſed *plus* ; the fire paſſing into it from the wire, and the bottle at the ſame time ſupplied from your hand. A ſuſpended ſmall cork-ball will play between theſe books 'till the equilibrium is reſtored.

EXPERIMENT IX.

When a body is electriſed *plus*, it will repel an electriſed feather, or ſmall cork-ball. When *minus* (or when in the common ſtate) it will attract them, but ſtronger when *minus* than when in the common ſtate, the difference being greater

EXPERIMENT X.

Though, as in *Experiment* VI. a man ſtanding on wax may be electriſed a number of times by repeatedly touching the wire of an electriſed bottle (held in the hand of one ſtanding on the floor) he receiving the fire from the wire each time : yet holding it in his own hand, and touching the wire, though he draws a ſtrong ſpark, and is violently ſhocked, no Electricity remains in him ; the fire only paſſing through him, from the upper to the lower part of the bottle. Obſerve, before the ſhock, to let ſome one on the floor touch him to reſtore the equilibrium in his body ; for in taking hold of the bottom of the bottle, he ſometimes becomes a little electriſed *minus*, which will continue after the ſhock, as would alſo any *plus* Electricity, which

he

he might have given him before the shock. For, restoring the equilibrium in the bottle, does not at all affect the E-lectricity in the man through whom the fire passes ; that Electricity is neither increased nor diminished.

EXPERIMENT XI.

The passing of the electrical fire from the upper to the lower part* of the bottle, to restore the equilibrium, is rendered strongly visible by the following pretty experiment. Take a book whose covering is filletted with gold ; bend a wire of eight or ten inches long, in the form of (*m*) Fig. 5. slip it on the end of the cover of the book, over the gold line, so as that the shoulder of it may press upon one end of the gold line, the ring up, but leaning towards the other end of the book. Lay the book on a glass or wax, and on the other end of the gold lines set the bottle electrised ; then bend the springing wire, by pressing it with a stick of wax till its ring approaches the ring of the bottle wire, instantly there is a strong spark and stroke, and the whole line of gold, which completes the communication, between the top and bottom of the bottle, will appear a vivid flame, like the sharpest lightning. The closer the contact between the shoulder of the wire, and the gold at one end of the line, and between the bottom of the bottle and the gold at the other end, the better the experiment succeeds. The room should be darkened. If you would

---

* *i. e.* from the *inside* to the *outside*.

have

have the whole filletting round the cover appear in fire at once, let the bottle and wire touch the gold in the diagonally oppoſite corners.

I am, &c.

B. FRANKLIN.

L E T T E R IV.

F R O M

BENJ. FRANKLIN, *Eſq;* in *Philadelphia,*

T O

PETER COLLINSON, Eſq; F. R. S. *London.*

Farther EXPERIMENTS *and* OBSERVATIONS *in*
ELECTRICITY.

SIR,   1748.

§ 1. THERE will be the ſame exploſion and ſhock if the electrified phial is held in one hand by the hook, and the coating touch'd with the other, as when held by the coating, and touch'd at the hook.

2. To

2. To take the charg'd phial fafely by the hook, and not at the fame time diminifh its force, it muft firft be fet down on an electric *per fe*.

3. The phial will be electrified as ftrongly, if held by the hook, and the coating apply'd to the globe or tube; as when held by the coating, and the hook apply'd *.

4. But the *direction* of the electrical fire being different in the charging, will alfo be different in the explofion. The bottle charged through the hook, will be difcharged through the hook; the bottle charged through the coating, will be difcharged through the coating, and not otherways; for the fire muft come out the fame way it went in.

5. To prove this, take two bottles that were equally charged through the hooks, one in each hand: bring their hooks near each other, and no fpark or fhock will follow; becaufe each hook is difpofed to give fire, and neither to receive it. Set one of the bottles down on glafs, take it up by the hook, and apply its coating to the hook of the other; then there will be an explofion and fhock, and both bottles will be difcharged.

6. Vary the experiment, by charging two phials equally, one through the hook, the other through the coating: hold that by the coating which was charged through the hook; and that by the hook which was charged through the coating: apply the hook of the firft to the coating of the

---

* This was a Difcovery of the very ingenious Mr *Kinnerfley's*, and by him communicated to me.

other,

other, and there will be no ſhock or ſpark. Set that down on glaſs which you held by the hook, take it up by the coating, and bring the two hooks together: a ſpark and ſhock will follow, and both phials be diſcharged.

In this experiment the bottles are totally diſcharged, or the equilibrium within them reſtored. The *abounding* of fire in one of the hooks (or rather in the internal ſurface of one bottle (being exactly equal to the *wanting* of the other: and therefore, as each bottle has in itſelf the *abounding* as well as the *wanting*, the wanting and abounding muſt be equal in each bottle. See §. 8, 9, 10, 11. But if a man holds in his hands two bottles, one fully electrified, the other not at all, and brings their hooks together, he has but half a ſhock, and the bottles will both remain half electrified, the one being half diſcharged, and the other half charged.

7. Place two phials equally charged on a table at five or ſix inches diſtance. Let a cork-ball, ſuſpended by a ſilk thread, hang between them. If the phials were both charged through their hooks, the cork, when it has been attracted and repelled by the one, will not be attracted, but equally repelled by the other. But if the phials were charged, the one through the hook, and the other * through the coating, the ball, when it is repelled from one hook,

* To charge a bottle commodiouſly through the coating, place it on a glaſs ſtand; form a communication from the prime conductor to the coating, and another from the hook to the wall or floor. When it is charged, remove the latter communication before you take hold of the bottle, otherwiſe great part of the fire will eſcape by it.

will

be as ftrongly attracted by the other, and play vigoroufly between them, till both phials are nearly difcharged.

8. When we ufe the terms of *charging* and *difcharging* the phial, it is in compliance with euftom, and for want of others more fuitable. Since we are of opinion that there is really no more electrical fire in the phial after what is called its *charging*, than before, nor lefs after its *difcharging*; excepting only the fmall fpark that might be given to, and taken from the non-electric matter, if feparated from the bottle, which fpark may not be equal to a five hundredth part of what is called the explofion.

For if, on the explofion, the electrical fire came out of the bottle by one part, and did not enter in again by another, then, if a man, ftanding on wax, and holding the bottle in one hand, takes the fpark by touching the wire hook with the other, the bottle being thereby *difcharged*, the man would be *charged*; or whatever fire was loft by one, would be found in the other, fince there was no way for its efcape : But the contrary is true.

9. Befides, the phial will not fuffer what is called a *charging*, unlefs as much fire can go out of it one way, as is thrown in by another. A phial cannot be charged ftanding on wax or glafs, or hanging on the prime conductor, unlefs a communication be formed between its coating and the floor.

10. But fufpend two or more phials on the prime conductor, one hanging to the tail of the other; and a wire from the laft to the floor, an equal number of turns of the

wheel

wheel fhall charge them all equally, and every one as much as one alone would have been. What is driven out at the tail of the firft, ferving to charge the fecond; what is driven out of the fecond charging the third; and fo on. By this means a great number of bottles might be charged with the fame labour, and equally high, with one alone, were it not that every bottle receives new fire, and lofes its old with fome relu&tance, or rather gives fome fmall re- fiftance to the charging, which in a number of botttles be- comes more equal to the charging power, and fo repels the fire back again on the globe, fooner than a fingle bottle would do.

11. When a bottle is charged in the common way, its *infide* and *outfide* furfaces ftand ready, the one to give fire by the hook, the other to receive it by the coating; the one is full, and ready to throw out, the other empty and extremely hungry; yet as the firft will not *give out*, unlefs the other can at the fame inftant *receive in*; fo neither will the latter receive in, unlefs the firft can at the fame inftant give out. When both can be done at once, it is done with inconceivable quicknefs and violence.

12. So a ftrait fpring (though the comparifon does not agree in every particular) when forcibly bent, muft, to re- ftore itfelf, contra&t that fide which in the bending was ex- tended, and extend that which was contra&ted; if either of thefe two operations be hindered, the other cannot be done. But the fpring is not faid to be *charg'd* with elafti-

E   city

city when bent, and difcharged when unbent ; its 'quantity of elafticity is always the fame.

13. Glafs, in like manner, has, within its fubftance, always the fame quantity of electrical fire, and that a very great quantity in proportion to the mafs of glafs, as fhall be fhewn hereafter.

14. This quantity, proportioned to the glafs, it ftrongly and obftinately retains, and will have neither more nor lefs though it will fuffer a change to be made in its parts and fituation ; *i. e.* we may take away part of it from one of the fides, provided we throw an equal quantity into the other.

15. Yet when the fituation of the electrical fire is thus altered in the glafs ; when fome has been taken from one fide, and fome added to the other, it will not be at reft or in its natural ftate, till it is reftored to its original equality.— And this reftitution cannot be made through the fubftance of the glafs, but muft be done by a non-electric communication formed without, from furface to furface.

16. Thus, the whole force of the bottle, and power of giving a fhock, is in the GLASS ITSELF ; the non-electrics in contact with the two furfaces, ferving only to *give* and *receive* to and from the feveral parts of the glafs ; that is, to give on one fide, and take away from the other.

17. This was difcovered here in the following manner : Purpofing to analyfe the electrified bottle, in order to find wherein its ftrength lay, we placed it on glafs, and drew out the cork and wire which for that purpofe had been

loofe-

loofely put in.   Then taking the bottle in one hand, and
bringing a finger of the other near its mouth, a ftrong fpark
came from the water, and the fhock was as violent as if the
wire had remained in it, which fhewed that the force did
not lie in the wire.   Then to find if it refided in the water,
being crouded into and condenfed in it, as confin'd by the
glafs, which had been our former opinion, we electrified
the bottle again, and placing it on glafs, drew out the wire
and cork as before ; then taking up the bottle, we decanted
all its water into an empty bottle, which likewife ftood on
glafs; and taking up that other bottle, we expected, if the
force refided in the water, to find a fhock from it ; but
there was none.   We judged then that it muft either be
loft in decanting, or remain in the firft bottle.   The latter
we found to be true ; for that bottle on trial gave the fhock,
though filled up as it ftood with frefh unelectrified water
from a tea-pot.—To find, then, whether glafs had this
property merely as glafs, or whether the form contributed
any thing to it; we took a pane of fafh-glafs, and laying
it on the hand, placed a plate of lead on its upper furface ;
then electrified that plate, and bringing a finger to it, there
was a fpark and fhock.   We then took two plates of lead
of equal dimenfions, but lefs than the glafs by two inches
every way, and electrified the glafs between them, by
electrifying the uppermoft lead ; then feparated the glafs
from the lead, in doing which, what little fire might
be in the lead was taken out, and the glafs being touched
in the electrified parts with a finger, afforded only very

fmall

fmall pricking fparks, but a great number of them might
be taken from different places. Then dexteroufly placing
it again between the leaden plates, and compleating a
circle between the two furfaces, a violent fhock enfued.——Which demonftrated the power to refide in
glafs as glafs, and that the non-electrics in contact ferved
only, like the armature of a loadftone, to unite the force
of the feveral parts, and bring them at once to any point
defired : it being the property of a non-electric, that the
whole body inftantly receives or gives what electrical fire is
given to or taken from any one of its parts.

18. Upon this we made what we called an *electrical-
battery*, confifting of eleven panes of large fafh-glafs, arm'd
with thin leaden plates, pafted on each fide, placed verti-
cally, and fupported at two inches diftance on filk cords,
with thick hooks of leaden wire, one from each fide,
ftanding upright, diftant from each other, and convenient
communications of wire and chain, from the giving fide of
one pane, to the receiving fide of the other ; that fo the
whole might be charged together, and with the fame labour as one fingle pane ; and another contrivance to bring
the giving fides, after charging, in contact with one long
wire, and the receivers with another, which two long
wires would give the force of all the plates of glafs at once
through the body of any animal forming the circle with
them. The plates may alfo be difcharged feparately, or
any number together that is required. But this machine
is not much ufed, as not perfectly anfwering our intention
with

with regard to the eafe of charging, for the reafon given, *Sec.* 10. We made alfo of large glafs panes, magical pictures, and felf-moving animated wheels, prefently to be defcribed.

19. I perceive by the ingenious Mr *Watfon*'s laft book, lately received, that Dr *Bevis* had ufed, before we had, panes of glafs to give a fhock *; though, till that book came to hand, I thought to have communicated it to you as a novelty. The excufe for mentioning it here is, that we tried the experiment differently, drew different confe-quences from it (for Mr *Watfon* ftill feems to think the fire *accumulated on the non-electric* that is in contact with the glafs, page 72) and, as far as we hitherto know, have carri-ed it farther.

20. The magical picture † is made thus. Having a large metzotinto with a frame and glafs, fuppofe of the KING, (God preferve him) take out the print, and cut a pannel out of it, near two inches diftant from the frame all round. If the cut is through the picture it is not the worfe. With thin pafte, or gum-water, fix the border that is cut off on the infide the glafs, preffing it fmooth and clofe ; then fill up the vacancy by gilding the glafs well with leaf gold, or brafs. Gild likewife the inner edge of the back of the frame all round, except the top part, and form a com-munication between that gilding and the gilding behind

* I have fince heard that Mr *Smeaton* was the firft who made ufe of panes of glafs for that purpofe.

† Contrived by Mr *Kinnerfley*.

the

the glaſs : then put in the board, and that ſide is finiſhed.
Turn up the glaſs, and gild the fore ſide exactly over the
back gilding, and when it is dry, cover it, by paſting on
the pannel of the picture that hath been cut out, obſerving
to bring the correſpondent parts of the border and picture
together, by which the picture will appear of a piece, as at
firſt, only part is behind the glaſs, and part before.—Hold
the picture horizontally by the top, and place a little
moveable gilt crown on the king's head. If now the picture be moderately electrified, and another perſon take hold
of the frame with one hand, ſo that his fingers touch its
inſide gilding, and with the other hand endeavour to take
off the crown, he will receive a terrible blow, and fail in
the attempt. If the picture were highly charged, the conſequence might perhaps be as fatal * as that of high treaſon,
for when the ſpark is taken through a quire of paper laid
on the picture, by means of a wire communication, it
makes a fair hole through every ſheet, that is, through for-
ty-eight leaves, (though a quire of paper is thought good
armour againſt the puſh of a ſword, or even againſt a piſ-
tol bullet, and the crack is exceeding loud. The operator,
who holds the picture by the upper end, where the inſide
of the frame is not gilt, to prevent its falling, feels nothing
of the ſhock, and may touch the face of the picture without danger, which he pretends is a teſt of his loyalty.—If a

* We have ſince found it fatal to ſmall animals, though not to large
ones.   The biggeſt we have yet killed is a hen.   1750.

ring

ring of perſons take the ſhock among them, the experiment is called, *The Conſpirators.*

21. On the principle, in *Sec.* 7, that hooks of bottles, differently charged, will attract and repel differently, is made an electrical wheel, that turns with conſiderable ſtrength. A ſmall upright ſhaft of wood paſſes at right angles through a thin round board, of about twelve inches diameter, and turns on a ſharp point of iron, fixed in the lower end, while a ſtrong wire in the upper end, paſſing through a ſmall hole in a thin braſs plate, keeps the ſhaft truly vertical. About thirty *radii* of equal length, made of ſaſh glaſs, cut in narrow ſtrips, iſſue horizontally from the circumference of the board, the ends moſt diſtant from the center being about four inches apart. On the end of every one, a braſs thimble is fixed. If now the wire of a bottle electrified in the common way, be brought near the circumference of this wheel, it will attract the neareſt thimble, and ſo put the wheel in motion ; that thimble, in paſſing by, receives a ſpark, and thereby being electrified is repelled, and ſo driven forwards ; while a ſecond being attracted, approaches the wire, receives a ſpark, and is driven after the firſt, and ſo on till the wheel has gone once round, when the thimbles before electrified approaching the wire, inſtead of being attracted as they were at firſt, are repelled, and the motion preſently ceaſes.—But if another bottle, which had been charged through the coating, be placed near the ſame wheel, its wire will attract the thimble repelled by the firſt, and thereby double the force that carries the wheel

round ;

round ; and not only taking out the fire that had been communicated to the thimbles by the firft bottle, but even robbing them of their natural quantity, inftead of being repelled when they come again towards the firft bottle, **they** are more ftrongly attracted, fo that the wheel mends its pace, till it goes with great rapidity twelve or fifteen rounds in a minute, and with fuch ftrength, as that the weight of one hundred *Spanifh* dollars with which we once loaded it, did not feem in the leaft to retard its motion.—This is called an electrical jack ; and if a large fowl were fpitted on the upright fhaft, it would be carried round before a fire with a motion fit for roafting.

22. But this wheel, like thofe driven by wind, water, or weights, moves by a foreign force, to wit, that of the bottles. The felf-moving wheel, though conftructed on the fame principles, appears more furprifing. 'Tis made of a thin round plate of window glafs, feventeen inches diameter, well gilt on both fides, all but two inches next the edge. Two fmall hemifpheres of wood are then fixed with cement to the middle of the upper and under fides, centrally oppofite, and in each of them a thick ftrong wire eight or ten inches long, which together make the axis of the wheel. It turns horizontally on a point at the lower end of its axis, which refts on a bit of brafs cemented within a glafs falt-cellar. The upper end of its axis paffes through a hole in a thin brafs plate cemented to a long ftrong piece of glafs, which keeps it fix or eight inches diftant from any non-electric, and has a fmall ball of wax or

metal

metal on its top to keep in the fire. In a circle on the table which fupports the wheel, are fixed twelve fmall pillars of glafs, at about four inches diftance, with a thimble on the top of each. On the edge of the wheel is a fmall leaden bullet, communicating by a wire with the gilding of the *upper* furface of the wheel ; and about fix inches from it is another bullet communicating in like manner with the *under* furface. When the wheel is to be charged by the upper furface, a communication muft be made from the under furface to the table. When it is well charged it begins to move ; the bullet neareft to a pillar moves towards the thimble on that pillar, and paffing by, electrifies it, and then pufhes itfelf from it; the fucceeding bullet, which communicates with the other furface of the glafs, more ftrongly attracts that thimble, on account of its being before electrified by the other bullet; and thus the wheel encreafes its motion till it comes to fuch a height as that the refiftance of the air regulates it. It will go half an hour, and make one minute with another twenty turns in a minute, which is fix hundred turns in the whole; the bullet of the upper furface giving in each turn twelve fparks, to the thimbles, which makes feven thoufand two hundred fparks ; and the bullet of the under furface receiving as many from the thimbles ; thofe bullets moving in the time near two thoufand five hundred feet —The thimbles are well fixed, and in fo exact a circle, that the bullets may pafs within a very fmall diftance of each of them.—If inftead of two bullets you put eight, four communi-

F

municating with the upper furface, and four with the under furface, placed alternately ; which eight, at about fix inches diftance, completes the circumference, the force and fwiftnefs will be greatly increafed, the wheel making fifty turns in a minute ; but then it will not continue moving fo long.——Thefe wheels may be applied, perhaps, to the ringing of chimes *, and moving of light-made orreries.

23. A fmall wire bent circularly, with a loop at each end ; let one end reft againft the under furface of the wheel, and bring the other end near the upper furface, it will give a terrible crack, and the force will be difcharged.

24. Every fpark in that manner drawn from the furface of the wheel, makes a round hole in the gilding, tearing off a part of it in coming out ; which fhews that the fire is not accumulated on the gilding, but is in the glafs itfelf.

25. The gilding being varnifhed over with turpentine varnifh, the varnifh, though dry and hard, is burnt by the fpark drawn through it, and gives a ftrong fmell and vifible fmoke. And when the fpark is drawn through paper, all round the hole made by it, the paper will be blacked by the fmoke, which fometimes penetrates feveral of the leaves. Part of the gilding torn off, is alfo found forcibly driven into the hole made in the paper by the ftroke.

* This was afterwards done with fuccefs by Mr *Kinnerfley.*

26. It

26. It is amazing to obferve in how fmall a portion of glafs a great electrical force may lie. A thin glafs bubble about an inch diameter, weighing only fix grains, being half filled with water, partly gilt on the outfide, and furnifh'd with a wire hook, gives, when electrified, as great a fhock as a man can well bear.   As the glafs is thickeft near the orifice, I fuppofe the lower half, which being gilt was electrified and gave the fhock, did not exceed two grains ; for it appeared, when broke, much thinner than the upper half.—If one of thefe thin bottles be electrified by the coating, and the fpark taken out through the gilding, it will break the glafs inwards, at the fame time that it breaks the gilding outwards.

27. And allowing (for- the reafons before given,·§. 8, 9, 10.) that there is no more electrical fire in a bottle after charging, than before, how great muft be the quantity in this fmall portion of glafs ! It feems as if it were of its very fubftance and effence. Perhaps if that due quantity of electrical fire fo obftinately retained by glafs, could be feparated from it, it would no longer be glafs; it might lofe its tranfparency, or its brittlenefs, or its elafticity.— Experiments may poffibly be invented hereafter, to difcover this.

27. We were furprifed at the account given in Mr *Watfon*'s book, of a fhock communicated through a great fpace of dry ground, and fufpect there muft be fome metalline quality in the gravel of that ground ; having found that

fimple

simple dry earth, rammed in a glass tube, open at both ends, and a wire hook inserted in the earth at each end, the earth and wires making part of a circle, would not conduct the least perceptible shock, and indeed when one wire was electrified, the other hardly showed any signs of its being in connection with it *. Even a thoroughly wet pack-thread sometimes fails of conducting a shock, though it otherwise conducts Electricity very well. A dry cake of ice, or an icicle held between two in a circle, likewise prevents the shock, which one would not expect, as water conducts it so perfectly well.—Gilding on a new book, though at first it conducts the shock extremely well, yet fails after ten or a dozen experiments, though it appears otherwise in all respects the same, which we cannot account for †.

28. There is one experiment more which surprizes us, and is not hitherto satisfactorily accounted for ; it is this : Place an iron shot on a glass stand, and let a ball of damp cork, suspended by a silk thread, hang in contact with the shot. Take a bottle in each hand, one that is electrified through the hook, the other through the coating : Apply the giving wire to the shot, which will electrify it *positive-*

* Probably the ground is never so dry.

† We afterwards found that it failed after one stroke with a large bottle ; and the continuity of the gold appearing broken, and many of its parts dissipated, the Electricity could not pass the remaining parts without leaping from part to part through the air, which always resists the motion of this fluid, and was probably the cause of the gold's not conducting so well as before.

ly,

ly, and the cork fhall be repelled: then apply the requiring wire, which will take out the fpark given by the other; when the cork will return to the fhot: Apply the fame again, and take out another fpark, fo will the fhot be electrified *negatively*, and the cork in that cafe fhall be repelled equally as before. Then apply the giving wire to the fhot, and give the fpark it wanted, fo will the cork return: Give it another, which will be an addition to its natural quantity, fo will the cork be repelled again: And fo may the experiment be repeated as long as there is any charge in the bottles. Which fhews that bodies having lefs than the common quantity of Electricity, repel each o-ther, as well as thofe that have more.

Chagrined a little that we have been hitherto able to produce nothing in this way of ufe to mankind; and the hot weather coming on, when electrical experiments are not fo agreeable, it is propofed to put an end to them for this feafon, fomewhat humoroufly, in a party of pleafure, on the banks of *Skuylkil* *. Spirits, at the fame time, are to be fired by a fpark fent from fide to fide through the river, without any other conductor than the water; an experiment which we fome time fince performed, to the amazement of many †. A turkey is to be killed for our

dinner

---

* The river that wafhes one fide of *Philadelphia*, as the *Delaware* does the other; both are ornamented with the fummer habitations of the citizens, and the agreeable manfions of the principal people of this colony.

† As the poffibility of this experiment has not been eafily conceived, I

fhall

dinner by the *electrical shock*, and roasted by the *electrical jack*, before a fire kindled by the *electrified bottle* : when the healths of all the famous electricians in *England, Holland, France*, and *Germany*, are to be drank in * *electrified bumpers*, under the discharge of guns from the *electrical battery*.

shall here describe it.—Two iron rods, about three feet long, were planted just within the margin of the river, on the opposite sides. A thick piece of wire, with a small round knob at its end, was fixed to the top of one of the rods, bending downwards, so as to deliver commodiously the spark upon the surface of the spirit. A small wire fastened by one end to the handle of the spoon, containing the spirit, was carried a-cross the river, and supported in the air by the rope commonly used to hold by, in drawing the ferry-boats over. The other end of this wire was tied round the coating of the bottle ; which being charged, the spark was delivered from the hook to the top of the rod standing in the water on that side. At the same instant the rod on the other side delivered a spark into the spoon, and fired the spirit. The electric fire returning to the coating of the bottle, through the handle of the spoon and the supported wire connected with them.

That the electric fire thus actually passes through the water, has since been satisfactorily demonstrated to many by an experiment of Mr *Kinnersley's*, performed in a trough of water about ten feet long. The hand being placed under water in the direction of the spark (which always takes the strait or shortest course) is struck and penetrated by it as it passes.

* An *electrified bumper* is a small thin glass tumbler, near filled with wine, and electrified as the bottle. This when brought to the ·lips gives a shock, if the party be close shaved, and does not breathe on the liquor.

April 29,
1749.

LET-

L E T T E R V.

CONTAINING

OBSERVATIONS *and* SUPPOSITIONS, *towards forming a new* HYPOTHESIS, *for explaining the several* Phænomena *of* THUNDER-GUSTS *.

S I R,

§. 1. NON-ELECTRIC bodies, that have electric fire thrown into them, will retain it till other non-electrics, that have lefs, approach ; and then it is communicated by a fnap, and becomes equally divided.

2. Electrical fire loves water, is ftrongly attracted by it, and they can fubfift together.

3. Air is an electric *per fe*, and when dry will not conduct the electrical fire ; it will neither receive it, nor give it to other bodies ; otherwife no body furrounded by air, could be electrified pofitively and negatively : for fhould it

---

* Thunder-gufts are fudden ftorms of thunder and lightning, which are frequently of fhort duration, but fometimes produce mifchievous effects.

be

be attempted pofitively: the air would immediately take a-
way the overplus ; or negatively, the air would fupply
what was wanting.

4. Water being electrified, the vapours arifing from it
will be equally electrified ; and floating in the air, in the
form of clouds, or otherwife, will retain that quantity of
electrical fire, till they meet with other clouds or bodies
not fo much electrified, and then will communicate as
before mentioned.

5. Every particle of matter electrified is repelled by
every other particle equally electrified. Thus the ftream
of a fountain, naturally denfe and continual, when electrified, will feparate and fpread in the form of a brufh, every
drop endeavouring to recede from every other drop. But
on taking out the electrical fire they clofe again.

6. Water being ftrongly electrified (as well as when
heated by common fire) rifes in vapours more copioufly ;
the attraction of cohefion among its particles being greatly
weakened, by the oppofite power of repulfion introduced
with the electrical fire ; and when any particle is by any
means difengaged, it is immediately repelled, and fo flies
into the air.

7. Particles happening to be fituated as *A* and *B*, (Fig.
VI. *reprefenting the profile of a veffel of water*) are more
eafily difengaged than *C* and *D*, as each is held by contact
with three only, whereas *C* and *D* are each in contact with
nine. When the furface of the water has the leaft motion,
parti-

particles are continually pufhed into the fituation reprefented by *A* and *B*.

8. Friction between a non-electric and an electric *per fe*
will produce electrical fire ; not by *creating*, but *collecting*
it : for it is equally diffufed in our walls, floors, earth, and
the whole mafs of common matter.   Thus the whirling
glafs globe, during its friction againft the cufhion, draws
fire from the cufhion, the cufhion is fupplied from the
frame of the machine, that from the floor on which it
ftands.   Cut off the communication by thick glafs or wax,
placed under the cufhion, and no fire can be *produced*, becaufe it cannot be *collected*.

9. The ocean is a compound of water, a non-electric,
and falt an electric *per fe*.

10. When there is a friction among the parts near its
furface, the electrical fire is collected from the parts below.
It is then plainly vifible in the night ; it appears at the
ftern and in the wake of every failing veffel ; every dafh
of an oar fhews it, and every furf and fpray : In ftorms the
whole fea feems on fire.—The detach'd particles of water
then repelled from the electrified furface, continually carry
off the fire as it is collected ; they rife and form clouds,
and thofe clouds are highly electrified, and retain the fire
till they have an opportunity of communicating it.

11 The particles of water rifing in vapours, attach themfelves to particles of air.

12. The particles of air are faid to be hard, round, feparate and diftant from each other ; every particle ftrongly

G   repelling

repelling every other particle, whereby they recede from each other, as far as common gravity will permit.

13. The space between any three particles equally repelling each other, will be an equilateral triangle.

14. In air compreffed, thefe triangles are fmaller; in rarified air they are larger.

15. Common fire joined with air, increafes the repulfion, enlarges the triangles, and thereby makes the air fpecifically lighter. Such air, among denfer air, will rife.

16. Common fire, as well as electrical fire, gives repulfion to the particles of water, and deftroys their attraction of cohefion; hence common fire, as well as electrical fire, affifts in raifing vapours.

17. Particles of water, having no fire in them, mutually attract each other. Three particles of water then being attached to the three particles of a triangle of air, would by their mutual attraction operating againft the air's repulfion, fhorten the fides and leffen the triangle, whereby that portion of air being made denfer, would fink to the earth with its water, and not rife to contribute to the formation of a cloud.

18. But if every particle of water attaching itfelf to air, brings with it a particle of common fire, the repulfion of the air being affifted and ftrengthened by the fire, more than obftructed by the mutual attraction of the particles of water, the triangle dilates, and that portion of air becoming rarer and fpecifically lighter rifes.

19. If the particles of water bring electrical fire when
they

they attach themfelves to air, the repulfion between the particles of water electrified, joins with the natural repulfion of the air, to force its particles to a greater diftance, whereby the triangles are dilated, and the air rifes, carrying up with it the water.

20. If the particles of water bring with them portions of *both forts* of fire, the repulfion of the particles of air is ftill more ftrengthened and increafed, and the triangles farther enlarged.

21. One particle of air may be furrounded by twelve particles of water of equal fize with itfelf, all in contact with it ; and by more added to thofe.

22. Particles of air thus loaded would be drawn nearer together by the mutual attraction of the particles of water, did not the fire, common or electrical, affift their repulfion.

23. If air thus loaded be compreffed by adverfe winds, or by being driven againft mountains, &c. or condenfed by taking away the fire that affifted it in expanding; the triangles contract, the air with its water will defcend as a dew; or, if the water furrounding one particle of air comes in contact with the water furrounding another, they coalefce and form a drop, and we have rain.

24. The fun fupplies (or feems to fupply) common fire to all vapours, whether raifed from earth or fea.

25. Thofe vapours which have both common and electrical fire in them, are better fupported, than thofe which have only common fire in them, For when vapours rife

into the coldeſt region above the earth, the cold will not diminiſh the electrical fire, if it doth the common.

26. Hence clouds formed by vapours raiſed from freſh waters within land, from growing vegetables, moiſt earth, &c. more ſpeedily and eaſily depoſite their water, having but little electrical fire to repel and keep the particles ſeparate. So that the greateſt part of the water raiſed from the land, is let fall on the land again ; and winds blowing from the land to the ſea are dry ; there being little uſe for rain on the ſea, and to rob the land of its moiſture, in order to rain on the ſea, would not appear reaſonable.

27. But clouds formed by vapours raiſed from the ſea, having both fires, and particularly a great quantity of the electrical, ſupport their water ſtrongly, raiſe it high, and being moved by winds, may bring it over the middle of the broadeſt continent from the middle of the wideſt ocean.

28. How theſe ocean clouds, ſo ſtrongly ſupporting their water, are made to depoſite it on the land where it is wanted, is next to be conſidered.

29. If they are driven by winds againſt mountains, thoſe mountains being leſs electrified attract them, and on contact take away their electrical fire (and being cold, the common fire alſo ;) hence the particles cloſe towards the mountains and towards each other. If the air was not much loaded, it only falls in dews on the mountain tops and ſides, forms ſprings, and deſcends to the vales in rivulets, which united, make larger ſtreams and rivers. If much loaded, the electrical fire is at once taken from the
whole

whole cloud ; and, in leaving it, flaſhes brightly and cracks loudly ; the particles inſtantly coaleſcing for want of that fire, and falling in a heavy ſhower.

30. When a ridge of mountains thus dams the clouds, and draws the electrical fire from the cloud firſt approaching it ; that which next follows, when it comes near the firſt cloud, now deprived of its fire, flaſhes into it, and begins to depoſite its own water ; the firſt cloud again flaſhing into the mountains ; the third approaching cloud, and all the ſucceeding ones, acting in the ſame manner as far back as they extend, which may be over many hundred miles of country.

31. Hence the continual ſtorms of rain, thunder, and lightning on the eaſt ſide of the *Andes*, which running north and ſouth, and being vaſtly high, intercept all the clouds brought againſt them from the *Atlantic* ocean by the trade winds, and oblige them to depoſite their waters, by which the vaſt rivers *Amazons*, *La Plata*, and *Oroonoko* are formed, which return the water into the ſame ſea, after having fertilized a country of very great extent.

32. If a country be plain, having no mountains to intercept the electrified clouds, yet it is not without means to make them depoſite their water. For if an electrified cloud coming from the ſea, meets in the air a cloud raiſed from the land, and therefore not electrified ; the firſt will flaſh its fire into the latter, and thereby both clouds ſhall be made ſuddenly to depoſite water.

33. The electrified particles of the firſt cloud cloſe when they loſe their fire ; the particles of the other cloud

cloſe

clofe in receiving it : in both, they have thereby an opportunity of coalefcing into drops.—The concuffion or jerk
given to the air, contributes alfo to fhake down the water,
not only from thofe two clouds, but from others near them.
Hence the fudden fall.of rain immediately after flafhes of
lightning.

34. To fhew this by an eafy experiment: Take two
round pieces of pafteboard two inches diameter; from the
center and circumference of each of them fufpend by
fine filk threads eighteen inches long, feven fmall balls of
wood, or feven peas equal in bignefs : fo will the balls ap-
pending to each pafteboard, form equal equilateral triangles, one ball being in the center, and fix at equal diftances
from that, and from each other ; and thus they reprefent
particles of air. Dip both fets in water, and fome adhering
to each ball, they will reprefent air loaded. Dexteroufly
electrify one fet, and its balls will repel each other to a
greater diftance, enlarging the triangles. Could the water
fupported by the feven balls come into contact, it would
form a drop or drops fo heavy as to break the cohefion
it had with the balls, and fo fall. Let the two fets then
reprefent two clouds, the one a fea cloud electrified,
the other a land cloud. Bring them within the fphere of
attraction, and they will draw towards each other,
and you will fee the feparated balls clofe thus; the firft
electrified ball that comes near an unelectrified ball by
attraction joins it, and gives it fire ; inftantly they feparate, and each flies to another ball of its own party, one to
give

give, the other to receive fire; and fo it proceeds through both fets, but fo quick as to be in a manner inftantaneous. In the collifion they fhake off and drop their water, which reprefents rain.

35. Thus when fea and land clouds would pafs at too great a diftance from the flafh, they are attracted towards each other till within that diftance ; for the fphere of e-lectrical attraction is far beyond the diftance of flafhing.

36. When a great number of clouds from the fea meet a number of clouds raifed from the land, the electrical flafhes appear to ftrike in different parts ; and as the clouds are joftled and mixed by the winds, or brought near by the electrical attraction, they continue to give and receive flafh after flafh, till the electrical fire is equally diffufed.

37. When the gun-barrel (in electrical experiments) has but little electrical fire in it, you muft approach it very near with your knuckle, before you can draw a fpark. Give it more fire, and it will give a fpark at a greater dif-tance. Two gun-barrels united, and as highly electrified, will give a fpark at a ftill greater diftance. But if two gun-barrels electrified will ftrike at two inches diftance, and make a loud fnap, to what a great diftance may 10,000 acres of electrified cloud ftrike and give its fire, and how loud muft be that crack ?

38. It is a common thing to fee clouds at different heights paffing different ways, which fhews different cur-rents of air, one under the other. As the air between the

tropics

tropics is rarified by the fun, it rifes, the denfer northern and fouthern air preffing into its place. The air fo rarified and forced up, paffes northward and fouthward, and muft defcend in the polar regions, if it has no opportunity before, that the circulation may be carried on.

39. As currents of air, with the clouds therein, pafs different ways, 'tis eafy to conceive how the clouds, paffing over each other, may attract each other, and fo come near enough for the electrical ftroke. And alfo how electrical clouds may be carried within land very far from the fea, before they have an opportunity to ftrike.

40. When the air, with its vapours raifed from the ocean between the tropics, comes to defcend in the polar regions, and to be in contact with the vapours arifing there, the electrical fire they brought begins to be communicated, and is feen in clear nights, being firft vifible where 'tis firft in motion, that is, where the contact begins, or in the moft northern part; from thence the ftreams of light feem to fhoot foutherly, even up to the zenith of northern countries. But tho' the light feems to fhoot from the north foutherly, the progrefs of the fire is really from the fouth northerly, its motion beginning in the north being the reafon that 'tis there firft feen.

For the electrical fire is never vifible but when in motion, and leaping from body to body, or from particle to particle thro' the air. When it paffes thro' denfe bodies 'tis unfeen. When a wire makes part of the circle, in the explofion of the electrical phial, the fire, though in great

quantity

quantity, paſſes in the wire inviſibly : but in paſſing along a chain, it becomes viſible as it leaps from link to link. In paſſing along leaf gilding 'tis viſible : for the leaf-gold is full of pores; hold a leaf to the light and it appears like a net, and the fire is ſeen in its leaping over the vacancies.—— And as when a long canal filled with ſtill water is opened at one end, in order to be diſcharged, the motion of the water begins firſt near the opened end, and proceeds towards the cloſe end, tho' the water itſelf moves from the cloſe towards the opened end : ſo the electrical fire diſcharged into the polar regions, perhaps from a thouſand leagues length of vaporiſed air, appears firſt where 'tis firſt in motion, *i. e.* in the moſt northern part, and the appearance proceeds ſouthward, tho' the fire really moves northward. This is ſuppoſed to account for the *Aurora Borealis.*

41. When there is great heat on the land, in a particular region (the ſun having ſhone on it perhaps ſeveral days, while the ſurrounding countries have been ſcreen'd by clouds) the lower air is rarified and riſes, the cooler denſer air above deſcends; the clouds in that air meet from all ſides, and join over the heated place; and if ſome are electrified, others not, lightning and thunder ſucceed, and ſhowers fall. Hence thunder-guſts after heats, and cool air after guſts; the water and the clouds that bring it, coming from a higher and therefore a cooler region.

42. An electrical ſpark, drawn from an irregular body at ſome diſtance is ſcarce ever ſtrait, but ſhows crooked

H

and

and waving in the air. So do the flashes of lightning; the clouds being very irregular bodies.

43. As electrified clouds pass over a country, high hills and high trees, lofty towers, spires, masts of ships, chimneys, &c. as so many prominencies and points, draw the electrical fire, and the whole cloud discharges there.

44. Dangerous, therefore, is it to take shelter under a tree, during a thunder-guft. It has been fatal to many, both men and beasts.

45. It is safer to be in the open field for another reason. When the cloaths are wet, if a flash in its way to the ground should strike your head, it may run in the water over the surface of your body; whereas, if your cloaths were dry, it would go through the body.

Hence a wet rat cannot be killed by the exploding electrical bottle, when a dry rat may *

46. Common fire is in all bodies, more or less, as well as electrical fire. Perhaps they may be different modifications of the same element; or they may be different elements. The latter is by some suspected.

47. If they are different things, yet they may and do subsist together in the same body.

48. When electrical fire strikes through a body, it acts upon the common fire contained in it, and puts that fire in motion; and if there be a sufficient quantity of each kind of fire, the body will be inflamed.

* This was tried with a bottle, containing about a quart. It is since thought that one of the large glass jars, mentioned in these papers, might have killed him, though wet.

49. When

49. When the quantity of common fire in the body is fmall, the quantity of the electrical fire (or the electrical ftroke) fhould be greater : if the quantity of common fire be great, lefs electrical fire fuffices to produce the effect.

50. Thus fpirits muft be heated before we can fire them by the electrical fpark*. If they are much heated, a fmall fpark will do ; if not, the fpark muft be greater.

51. 'Till lately we could only fire warm vapours ; but now we can burn hard dry rofin. And when we can procure greater electrical fparks, we may be able to fire not only unwarm'd fpirits, as lightning does, but even wood, by giving fufficient agitation to the common fire contained in it, as friction we know will do.

52. Sulphureous and inflammable vapours arifing from the earth, are eafily kindled by lightning. Befides what arife from the earth, fuch vapours are fent out by ftacks of moift hay, corn, or other vegetables, which heat and reek. Wood rotting in old trees or buildings does the fame. Such are therefore eafily and often fired.

53. Metals are often melted by lightning, tho' perhaps not from heat in the lightning, nor altogether from agitated fire in the metals.—For as whatever body can infinuate itfelf between the particles of metal, and overcome the attraction by which they cohere (as fundry menftrua

---

* We have fince fired fpirits without heating them, when the weather is warm. A little poured into the palm of the hand, will be warmed fufficiently by the hand, if the fpirit be well rectified. Æther takes fire moft readily.

can) will make the folid become a fluid, as well as fire, yet without heating it : fo the electrical fire, or lightning, creating a violent repulfion between the particles of the metal it paffes through, the metal is fufed.

54. If you would, by a violent fire, melt off the end of a nail, which is half driven into a door, the heat given the whole nail before a part would melt, muft burn the board it fticks in. And the melted part would burn the floor it dropp'd on. But if a fword can be melted in the fcabbard, and money in a man's pocket, by lightning, without burning either, it muft be a cold fufion *.

55. Lightning rends fome bodies. The electrical fpark will ftrike a hole through a quire of ftrong paper.

56. If the fource of lightning, affigned in this paper, be the true one, there fhould be little thunder heard at fea far from land. And accordingly fome old fea-captains, of whom enquiry has been made, do affirm, that the fact a-grees perfectly with the hypothefis ; for that in croffing the great ocean, they feldom meet with thunder till they come into foundings ; and that the iflands far from the continent have very little of it. And a curious obferver, who lived 13 years at *Bermudas*, fays, there was lefs thunder there in that whole time than he has fometimes heard in a month at *Carolina*.

* Thefe facts, though related in feveral accounts, are now doubted ; fince it has been obferved that the parts of a bell-wire which fell on the floor being broken and partly melted by lightning, did actually burn into the boards. (See *Philof. Tranf.* Vol. LI. Part I. and Mr *Kinnerfley* has found that a fine iron wire, melted by Electricity, has had the fame effect.)

A D-

ADDITIONAL PAPERS

TO

PETER COLLINSON, *Esq*; F.R.S. *London.*

S I R,   *Philadelphia, July* 29, 1750.

AS you firſt put us on electrical experiments, by
ſending to our library company a tube, with directions how to uſe it ; and as our honourable proprietary
enabled us to carry thoſe experiments to a greater height,
by his generous preſent of a compleat electrical apparatus ; tis fit that both ſhould know, from time to time,
what progreſs we make. It was in this view I wrote and
ſent you my former papers on this ſubject, deſiring, that
as I had not the honour of a direct correſpondence with
that bountiful benefactor to our library, they might be
communicated to him through your hands. In the ſame
view I write and ſend you this additional paper. If it
happens to bring you nothing new (which may well be,
conſidering the number of ingenious men in *Europe*, continually engaged in the ſame reſearches) at leaſt it will
ſhow, that the inſtruments put into our hands are not neglected ; and, that if no valuable diſcoveries are made by
us, whatever the cauſe may be, it is not want of induſtry
and application.

                I am, Sir,
                    Your much obliged
                    Humble Servant,

                        B. FRANKLIN.

OPINIONS *and* CONJECTURES,

*concerning the Properties and Effects of the
electrical Matter, arising from Experiments
and Observations, made at* Philadelphia, 1749.

§. 1. THE electrical matter consists of particles extremely subtile, since it can permeate common matter, even the densest metals, with such ease and freedom as not to receive any perceptible resistance.

2. If any one should doubt whether the electrical matter passes thro' the substance of bodies, or only over and along their surfaces, a shock from an electrified large glass jar, taken through his own body, will probably convince him.

3. Electrical matter differs from common matter in this, that the parts of the latter mutually attract, those of the former mutually repel, each other. Hence the appearing divergency in a stream of electrified effluvia.

4. But though the particles of electrical matter do repel each other, they are strongly attracted by all other matter *.

* See the ingenious essays on Electricity, in the *Transactions*, by Mr *Ellicot*.

5. From

5. From thefe three things, the extreme fubtilty of the electrical matter, the mutual repulfion of its parts, and the ftrong attraction between them and other matter, arife this effect, that, when a quantity of electrical matter is applied to a mafs of common matter, of any bignefs or length, within our obfervation (which hath not already got its quantity) it is immediately and equally diffufed through the whole.

6. Thus common matter is a kind of fpunge to the electrical fluid. And as a fpunge would receive no water if the parts of water were not fmaller than the pores of the fpunge ; and even then but flowly, if there were not a mutual attraction between thofe parts and the parts of the fpunge ; and would ftill imbibe it fafter, if the mutual attraction among the parts of the water did not impede, fome force being required to feparate them ; and fafteft, if, inftead of attraction, there were a mutual repulfion among thofe parts, which would act in conjunction with the attraction of the fpunge. So is the cafe between the electrical and common matter.

7. But in common matter there is (generally) as much of the electrical as it will contain within its fubftance. If more is added, it lies without upon the furface, and forms what we call an electrical atmofphere ; and then the body is faid to be electrified.

8. 'Tis fuppofed, that all kinds of common matter do not attract and retain the electrical, with equal ftrength and force, for reafons to be given hereafter. And that thofe

called

called electrics *per fe*, as glafs, &c. attract and retain it
ftrongeft, and contain the greateft quantity.

9. We know that the electrical fluid is *in* common
matter, becaufe we can pump it *out* by the globe or tube.
We know that common matter has near as much as it can
contain, becaufe, when we add a little more to any portion of it, the additional quantity does not enter, but forms
an electrical atmofphere. And we know that common
matter has not (generally) more than it can contain, otherwife all loofe portions of it would repel each other, as
they conftantly do when they have electric atmofpheres.

10. The beneficial ufes of this electric fluid in the
creation, we are not yet well acquainted with, though
doubtlefs fuch there are, and thofe very confiderable ; but
we may fee fome pernicious confequences that would attend a much greater proportion of it. For had this globe
we live on, as much of it in proportion as we can give to a
globe of iron, wood, or the like, the particles of duft and
other light matters that get loofe from it, would, by virtue
of their feparate electrical atmofpheres, not only repel each
other, but be repelled from the earth, and not eafily be
brought to unite with it again ; whence our air would
continually be more and more clogged with foreign matter, and grow unfit for refpiration. This affords another
occafion of adoring that wifdom which has made all things
by weight and meafure !

11. If a piece of common matter be fuppofed entirely
free from electrical matter, and a fingle particle of the
latter

latter be brought nigh, it will be attracted, and enter the body, and take place in the center, or where the attraction is every way equal. If more particles enter, they take their places where the balance is equal between the attraction of the common matter, and their own mutual repulsion. Tis suppofed they form triangles, whofe fides fhorten as their number increafes; 'till the common matter has drawn in fo many, that its whole power of compreffing thofe triangles by attraction, is equal to their whole power of expanding themfelves by repulfion; and then will fuch piece of matter receive no more.

12. When part of this natural proportion of electrical fluid is taken out of a piece of common matter, the triangles formed by the remainder, are fuppofed to widen by the mutual repulfion of the parts, until they occupy the whole piece.

13. When the quantity of electrical fluid, taken from a piece of common matter, is reftored again, it enters, the expanded triangles being again compreffed till there is room for the whole.

14. To explain this: take two apples, or two balls of wood or other matter, each having its own natural quantity of the electrical fluid. Sufpend them by filk lines from the cieling. Apply the wire of a well-charged vial, held in your hand, to one of them (A) *Fig.* 7, and it will receive from the wire a quantity of the electrical fluid; but will not imbibe it, being already full. The fluid therefore will flow round its furface, and form an electrical atmofphere.

I

Bring

Bring A into contact with B, and half the electrical fluid is communicated, so that each has now an electrical atmosphere, and therefore they repel each other. Take away these atmospheres by touching the balls, and leave them in their natural state : then, having fixed a stick of sealing-wax to the middle of the vial to hold it by, apply the wire to A, at the same time the coating touches B. Thus will a quantity of the electrical fluid be drawn out of B, and thrown on A. So that A will have a redundance of this fluid, which forms an atmosphere round it, and B an exactly equal deficiency. Now, bring these balls again into contact, and the electrical atmosphere will not be divided between A and B, into two smaller atmospheres as before ; for B will drink up the whole atmosphere of A, and both will be found again in their natural state.

15. The form of the electrical atmosphere is that of the body it surrounds. This shape may be rendered visible in a still air, by raising a smoke from dry rosin, dropt into a hot tea-spoon under the electrised body, which will be attracted, and spread itself equally on all sides, covering and concealing the body*. And this form it takes, because it is attracted by all parts of the surface of the body, though it cannot enter the substance already replete. Without this attraction, it would not remain round the body, but dissipate in the air.

* See pa e 6.

16. The

16. The atmofphere of electrical particles furrounding an electrified fphere, is not more difpofed to leave it, or more eafily drawn off from any one part of the fphere than from another, becaufe it is equally attracted by every part. But that is not the cafe with bodies of any other figure. From a cube it is more eafily drawn at the corners than at the plane fides, and fo from the angles of a body of any other form, and ftill moft eafily from the angle that is moft acute. Thus if a body fhaped as A,B,C,D,E, in Fig. 8. be electrified, or have an electrical atmofphere communicated to it, and we confider every fide as a bafe on which the particles reft, and by which they are attracted, one may fee, by imagining a line from A to F, and another from E to G, that the portion of the atmofphere included in F,A,E,G, has the line A E for its bafis. So the portion of atmofphere included in H, A, B, I, has the line A, B, for its bafis. And likewife the portion included in K, B, C, L, has B, C, to reft on ; and fo on the other fide of the figure. Now if you would draw off this atmofphere with any blunt fmooth body, and approach the middle of the fide A, B, you muft come very near, before the force of your attracter exceeds the force or power with which that fide holds its atmofphere. But there is a fmall portion between I, B, K, that has lefs of the furface to reft on, and to be attracted by, than the neighbouring portions, while at the fame time there is a mutual repulfion between its particles, and the particles of thofe portions, therefore here you can get it with more eafe, or at

a great-

a greater diſtance. Between F, A, H, there is a larger portion that has yet a leſs ſurface to reſt on, and to attract it;
here therefore you can get it away ſtill more eaſily. But
eaſieſt of all between L, C, M, where the quantity is largeſt,
and the ſurface to attract and keep it back the leaſt.
When you have drawn away one of theſe angular portions
of the fluid, another ſucceeds in its place, from the nature of fluidity and the mutual repulſion before-mentioned;
and ſo the atmoſphere continues flowing off at ſuch angle,
like a ſtream, till no more is remaining. The extremities
of the portions of atmoſphere over theſe angular parts, are
likewiſe at a greater diſtance from the electrified body, as
may be ſeen by the inſpection of the above figure; the
point of the atmoſphere of the angle C, being much farther from C, than any other part of the atmoſphere over
the lines C, B, or B, A: And, beſides the diſtance ariſing
from the nature of the figure, where the attraction is leſs,
the particles will naturally expand to a greater diſtance by
their mutual repulſion. On theſe accounts we ſuppoſe e-
lectrified bodies diſcharge their atmoſpheres upon unelectrified bodies more eaſily, and at a greater diſtance from
their angles and points than from their ſmooth ſides. —
Thoſe points will alſo diſcharge into the air, when the body has too great an electrical atmoſphere, without bringing any non-electric near, to receive what is thrown off:
For the air, though an electric *per ſe*, yet has always more
or leſs water and other non-electric matters mixed with it:
and theſe attract and receive what is ſo diſcharged.

17. But

17. But points have a property, by which they *draw on* as well as *throw off* the electrical fluid, at greater diftances than blunt bodies can.   That is, as the pointed part of an electrified body will difcharge the atmófphere of that body, or communicate it fartheft to another body, fo the point of an unelectrified body will draw off the electrical atmofphere from an electrified body, farther than a blunter part of the fame unelectrified body will do.   Thus a pin held by the head, and the point prefented to an electrified body, will draw off its atmofphere at a foot diftance; where, if the head were prefented inftead of the point, no fuch effect would follow.   To underftand this, we may confider, that if a perfon ftanding on the floor would draw off the electrical atmofphere from an electrified body, an iron crow and a blunt knitting-needle held alternately in his hand, and prefented for that purpofe, do not draw with different forces in proportion to their different maffes. For the man, and what he holds in his hand, be it large or fmall, are connected with the common mafs of un-electrified matter; and the force with which he draws is the fame in both cafes, it confifting in the different pro-portion of electricity in the electrified body, and that com-mon mafs.   But the force with which the electrified body retains its atmofphere by attracting it, is proportioned to the furface over which the particles are placed; *i. e.* four fquare inches of that furface retain their atmofphere with four times the force that one fquare inch retains its at-mofphere.   And as in plucking the hairs from the horfe's

tail, a degree of ftrength not fufficient to pull away a handful at once, could yet eafily ftrip it hair by hair ; fo a blunt body prefented cannot draw off a number of particles at once, but a pointed one, with no greater force, takes them away eafily, particle by particle.

18. Thefe explanations of the power and operation of points, when they firft occurr'd to me, and while they firft floated in my mind, appeared perfectly fatisfactory ; but now I have wrote them, and confidered them more clofely in black and white, I muft own I have fome doubts about them ; yet, as I have at prefent nothing better to offer in their ftead, I do not crofs them out : for even a bad folution read, and its faults difcovered, has often given rife to a good one, in the mind of an ingenious reader.

19. Nor is it of much importance to us, to know the manner in which nature executes her laws ; 'tis enough if we know the laws themfelves. 'Tis of real ufe to know that china left in the air unfupported will fall and break ; but *how* it comes to fall, and *why* it breaks, are matters of fpeculation. 'Tis a pleafure indeed to know them, but we can preferve our china without it.

20. Thus in the prefent cafe, to know this power of points, may poffibly be of fome ufe to mankind, though we fhould never be able to explain it. The following experiments, as well as thofe in my firft paper, fhew this power. I have a large prime conductor, made of feveral thin fheets of clothier's pafteboard, form'd into a tube, near

ten

ten feet long and a foot diameter. It is cover'd with *Dutch* embofs'd paper, almoſt totally gilt. This large metallic furface fupports a much greater electrical atmoſphere than a rod of iron of 50 times the weight would do. It is fufpended by filk lines, and when charged will ſtrike at near two inches diſtance, a pretty hard ſtroke, fo as to make ones knuckle ach. Let a perfon ſtanding on the floor prefent the point of a needle at 12 or more inches diſtance from it, and while the needle is fo prefented, the conductor cannot be charged, the point drawing off the fire as faſt as it is thrown on by the electrical globe. Let it be charged, and then prefent the point at the fame diſtance, and it will fuddenly be difcharged. In the dark you may fee a light on the point, when the experiment is made. And if the perfon holding the point ſtands upon wax, he will be electrified by receiving the fire at that diſtance. Attempt to draw off the electricity with a blunt body, as a bolt of iron round at the end, and fmooth (a filverfmith's iron punch, inch thick, is what I ufe) and you muſt bring it within the diſtance of three inches before you can do it, and then it is done with a ſtroke and crack. As the paſteboard tube hangs loofe on filk lines, when you approach it with the punch iron, it likewife will move towards the punch, being attracted while it is charged ; but if, at the fame inſtant, a point be prefented as before, it retires again, for the point difcharges it. Take a pair of large brafs fcales, of two or more feet beam, the cords of the fcales being filk. Sufpend

pend the beam by a pack-thread from the cieling, fo that the bottom of the fcales may be about a foot from the floor : The fcales will move round in a circle by the untwifting of the packthread. Set the iron punch on the end upon the floor, in fuch a place as that the fcales may pafs over it in making their circle : Then electrify one fcale, by applying the wire of a charged phial to it. As they move round, you fee that fcale draw nigher to the floor, and dip more when it comes over the punch; and if that be placed at a proper diftance, the fcale will fnap and difcharge its fire into it. But if a needle be ftuck on the end of the punch, its point upwards, the fcale, inftead of drawing nigh to the punch, and fnapping, difcharges its fire filently through the point, and rifes higher from the punch. Nay, even if the needle be placed upon the floor near the punch, its point upwards, the end of the punch, tho' fo much higher than the needle, will not attract the fcale and receive its fire, for the needle will get it and convey it away, before it comes nigh enough for the punch to act. And this is conftantly obfervable in thefe experiments, that the greater quantity of electricity on the pafteboard tube, the farther it ftrikes or difcharges its fire, and the point likewife will draw it off at a ftill greater diftance.

Now if the fire of electricity and that of lightning be the fame, as I have endeavoured to fhew at large, in a former paper, this pafteboard tube and thefe fcales may reprefent electrified clouds. If a tube of only ten feet
long

long will ftrike and difcharge its fire on the punch at two or three inches diftance, an electrified cloud of per- haps 10,000 acres may ftrike and difcharge on the earth at a proportionably greater diftance. The horizontal mo- tion of the fcales over the floor, may reprefent the mo- tion of the clouds over the earth; and the erect iron punch, a hill or high building; and then we fee how electrified clouds paffing over hills or high buildings at too great a height to ftrike, may be attracted lower till within their ftriking diftance. And laftly, if a needle fixed on the punch with its point upright, or even on the floor below the punch, will draw the fire from the fcale filently at a much greater than the ftriking diftance, and fo prevent its defcending towards the punch; or if in its courfe it would have come nigh enough to ftrike, yet be- ing firft deprived of its fire it cannot, and the punch is thereby fecured from the ftroke. I fay, if thefe things are fo, may not the knowledge of this power of points be of ufe to mankind, in preferving houfes, churches, fhips, &c. from the ftroke of lightning, by directing us to fix on the higheft parts of thofe edifices, upright rods of iron made fharp as a needle, and gilt to prevent ruft- ing, and from the foot of thofe rods a wire down the outfide of the building into the ground, or down round one of the fhrouds of a fhip, and down her fide till it reaches the water? Would not thefe pointed rods proba- bly draw the electrical fire filently out of a cloud before

K

it

it came nigh enough to ftrike, and thereby fecure us from that moft fudden and terrible mifchief?

21. To determine the queftion, whether the clouds that contain lightning are electrified or not, I would pro-pofe an experiment to be try'd where it may be done con-veniently. On the top of fome high tower or fteeple, place a kind of centry-box (as in FIG. 9.) big enough to contain a man and an electrical ftand. From the middle of the ftand let an iron rod rife and pafs bending out of the door, and then upright 20 or 30 feet, pointed very fharp at the end. If the electrical ftand be kept clean and dry, a man ftand-ing on it when fuch clouds are paffing low, might be elec-trified and afford fparks, the rod drawing fire to him from a cloud. If any danger to the man fhould be apprehended (though I think there would be none) let him ftand on the floor of his box, and now and then bring near to the rod the loop of a wire that has one end faftened to the leads, he holding it by a wax handle ; fo the fparks, if the rod is electrified, will ftrike from the rod to the wire, and not affect him.

22. Before I leave this fubject of lightning, I may men-tion fome other fimilarities between the effects of that, and thofe of electricity. Lightning has often been known to ftrike people blind. A pigeon that we ftruck dead to appearance by the electrical fhock, recovering life, drooped about the yard feveral days, eat nothing, though crumbs were thrown to it, but declined and died. We did not think of its being deprived of fight ; but afterwards a

pullet

pullet ſtruck dead in like manner, being recovered by repeatedly blowing into its lungs, when ſet down on the
floor, ran headlong againſt the wall, and on examination
appeared perfectly blind. Hence we concluded that the
pigeon alſo had been abſolutely blinded by the ſhock
The biggeſt animal we have yet killed, or tried to kill,
with the electrical ſtroke, was a well-grown pullet.

23. Reading in the ingenious Dr *Miles*'s account of
the thunder ſtorm at *Stretham*, the effect of the lightning
in ſtripping off all the paint that had covered a gilt
moulding of a pannel of wainſcot, without hurting the
reſt of the paint, I had a mind to lay a coat of paint over
the filletting of gold on the cover of a book, and try the
effect of a ſtrong electrical flaſh ſent through that gold
from a charged ſheet of glaſs. But having no paint at
hand, I paſted a narrow ſtrip of paper over it; and when
dry, ſent the flaſh through the gilding, by which the paper
was torn off from end to end, with ſuch force, that it was
broke in ſeveral places, and in others brought away part of
the grain of the Turky-leather in which it was bound;
and convinced me, that had it been painted, the paint
would have been ſtript off in the ſame manner with that
on the wainſcot at *Stretham*.

24. Lightning melts metals, and I hinted in my paper
on that ſubject, that I ſuſpected it to be a cold fuſion;
I do not mean a fuſion by force of cold, but a fuſion
without heat *. We have alſo melted gold, ſilver, and

* See note in page 49.

copper,

copper, in fmall quantities, by the electrical flafh. The manner is this: Take leaf gold, leaf filver, or leaf gilt copper, commonly called leaf brafs, or *Dutch* gold ; cut off from the leaf long narrow ftrips, the breadth of a ftraw. Place one of thefe ftrips between two ftrips of fmooth glafs that are about the width of your finger. If one ftrip of gold, the length of the leaf, be not long e-nough for the glafs, add another to the end of it, fo that you may have a little part hanging out loofe at each end of the glafs. Bind the pieces of glafs together from end to end with ftrong filk thread ; then place it fo as to be part of an electrical circuit, (the ends of gold hanging out being of ufe to join with the other parts of the circuit) and fend the flafh through it, from a large electrified jar or fheet of glafs. Then if your ftrips of glafs remain whole, you will fee that the gold is miffing in feveral places, and inftead of it a metallic ftain on both the glaffes ; the ftains on the upper and under glafs exactly fimilar in the minuteft ftroke, as may be feen by holding them to the light; the metal appeared to have been not only melted, but even vitrified, or otherwife fo driven into the pores of the glafs, as to be protected by it from the action of the ftrongeft *Aqua Fortis*, or *Aqua Regia*. I fend you enclofed two little pieces of glafs with thefe metallic ftains upon them, which cannot be removed with-out taking part of the glafs with them. Sometimes the ftain fpreads a little wider than the breadth of the leaf, and looks brighter at the edge, as by infpecting clofely

you

you may obſerve in theſe. Sometimes the glaſs breaks to pieces; once the upper glaſs broke into a thouſand pieces, looking like coarſe ſalt. Theſe pieces I ſend you were ſtain'd with *Dutch* gold. True gold makes a darker ſtain, ſomewhat reddiſh; ſilver, a greeniſh ſtain. We once took two pieces of thick looking-glaſs, as broad as a *Gunter's* ſcale, and ſix inches long; and placing leaf-gold between them, put them between two ſmoothly plain'd pieces of wood, and fix'd them tight in a book-binder's ſmall preſs; yet though they were ſo cloſely confined, the force of the electrical ſhock ſhivered the glaſs into many pieces. The gold was melted, and ſtain d into the glaſs, as uſual. The circumſtances of the breaking of the glaſs differ much in making the experiment, and ſometimes it does not break at all: but this is conſtant, that the ſtains in the upper and under pieces are exact counterparts of each other. And though I have taken up the pieces of glaſs between my fingers immediately after this melting, I never could perceive the leaſt warmth in them.

25. In one of my former papers, I mentioned, that, gilding on a book, though at firſt it communicated the ſhock perfectly well, yet failed after a few experiments, which we could not account for. We have ſince found that one ſtrong ſhock breaks the continuity of the gold in the filletting, and makes it look rather like duſt of gold, abundance of its parts being broken and driven off; and it will ſeldom conduct above one ſtrong ſhock. Perhaps this may be the reaſon : When there is not a perfect

continuity

continuity in the circuit, the fire muft leap over the vacancies : There is a certain diftance which it is able to
leap over according to its ftrength ; if a number of
fmall vacancies, though each be very minute, taken together exceed that diftance, it cannot leap over them, and
fo the fhock is prevented.

26. From the before-mentioned law of electricity, that
points as they are more or lefs acute, draw on and
throw off the electrical fluid with more or lefs power,
and at greater or lefs diftances, and in larger or fmaller
quantities in the fame time, we may fee how to account
for the fituation of the leaf of gold fufpended between
two plates, the upper one continually electrified, the under one in a perfon's hand ftanding on the floor.   When
the upper plate is electrified, the leaf is attracted, and
raifed towards it, and would fly to that plate, were it not
for its own points.   The corner that happens to be uppermoft when the leaf is rifing, being a fharp point, from
the extream thinnefs of the gold, draws and receives at
a diftance a fufficient quantity of the electric fluid to
give itfelf an electric atmofphere, by which its progrefs
to the upper plate is ftopt, and it begins to be repelled
from that plate, and would be driven back to the
under plate, but that its loweft corner is likewife a
point, and throws off or difcharges the overplus of the
leaf's atmofphere, as faft as the upper corner draws it
on.   Were thefe two points perfectly equal in acutenefs,
the leaf would take place exactly in the middle fpace,
for

for its weight is a trifle, compared to the power acting
on it : But it is generally neareſt the unelectrified plate,
becauſe, when the leaf is offered to the electrified plate,
at a diſtance, the ſharpeſt point is commonly firſt affected
and raiſed towards it ; ſo *that* point, from its greater
acuteneſs, receiving the fluid faſter than its oppoſite can
diſcharge it at equal diſtances, it retires from the electrified plate, and draws nearer to the unelectrified plate,
till it comes to a diſtance where the diſcharge can be exactly equal to the receipt, the latter being leſſened, and
the former encreaſed ; and there it remains as long as the
globe continues to ſupply freſh electrical matter. This
will appear plain, when the difference of acuteneſs in
the corners is made very great. Cut a piece of *Dutch*
gold (which is fitteſt for theſe experiments on account of
its greater ſtrength) into the form of FIG. 10. the upper corner a right angle, the two next obtuſe angles,
and the loweſt a very acute one ; and bring this on your
plate under the electrified plate, in ſuch a manner as
that the right-angled part may be firſt raiſed (which is
done by covering the acute part with the hollow of your
hand) and you will ſee this leaf take place much nearer
to the upper than the under plate ; becauſe without
being nearer, it cannot receive ſo faſt at its right-angled
point, as it can diſcharge at its acute one. Turn this
leaf with the acute part uppermoſt, and then it takes
place neareſt the unelectrified plate ; becauſe, otherwiſe,
it receives faſter at its acute point than it can diſcharge

at

at its right-angled one. Thus the difference of diftance is always proportioned to the difference of acutenefs. Take care in cutting your leaf, to leave no little ragged particles on the edges, which fometimes form points where you would not have them. You may make this figure fo acute below, and blunt above, as to need no under plate, it difcharging faft enough into the air. When it is made narrower, as the figure between the pricked lines we call it the *Golden Fifh*, from its manner of acting. For if you take it by the tail, and hold it at a foot or greater horizontal diftance from the prime conductor, it will, when let go, fly to it with a brifk but wavering motion, like that of an eel through the water; it will then take place under the prime conductor, at perhaps a quarter or half an inch diftance, and keep a continual fhaking of its tail like a fifh, fo that it feems animated. Turn its tail towards the prime conductor, and then it flies to your finger, and feems to nibble it. And if you hold a plate under it at fix or eight inches diftance, and ceafe turning the globe, when the electrical atmofphere of the conductor grows fmall, it will defcend to the plate and fwim back again feveral times with the fame fifh-like motion, greatly to the entertainment of fpectators. By a little practice in blunting or fharpening the heads or tails of thefe figures, you may make them take place as defired, nearer or farther from the electrified plate.

27. It is faid in Section 8, of this paper, that all kinds of common matter are fuppofed not to attract the electrical
fluid

fluid with equal ſtrength ; and that thoſe called electrics *per ſe,* as glaſs, *&c.* attract and retain it ſtrongeſt, and contain the greateſt quantity.   This latter poſition may ſeem a paradox to ſome, being contrary to the hitherto received opinion ; and therefore I ſhall now endeavour to explain it.

28. In order to this, let it firſt be conſider'd, *that we cannot by any means we are yet acquainted with, force the electrical fluid thro' glaſs.*   I know it is commonly thought that it eaſily prevades glaſs; and the experiment of a feather ſuſpended by a thread, in a bottle hermetically ſealed, yet moved by bringing a rubbed tube near the outſide of the bottle, is alledged to prove it.   But, if the electrical fluid ſo eaſily pervades glaſs, how does the vial become *charged* (as we term it) when we hold it in our hands ?   Would not the fire thrown in by the wire, paſs through to our hands, and ſo eſcape into the floor ?   Would not the bottle in that caſe be left juſt as we found it, uncharged, as we know a metal bottle ſo attempted to be charged would be ?   Indeed, if there be the leaſt crack, the minuteſt ſolution of continuity in the glaſs, though it remains ſo tight that nothing elſe we know of will paſs, yet the extremely ſubtile electric fluid flies through ſuch a crack with the greateſt freedom, and ſuch a bottle we know can never be charged : What then makes the difference between ſuch a bottle and one that is ſound, but this, that the fluid can paſs through the one, and not through the other * ?

* See the firſt ſixteen Sections of the former paper, called *Farther Experiments,* &c.

L   29. It

29. It is true, there is an experiment that at firſt ſight would be apt to ſatisfy a ſlight obſerver, that the fire thrown into the bottle by the wire, does really paſs thro' the glaſs. It is this : place the bottle on a glaſs ſtand, un-der the prime conductor ; ſuſpend a bullet by a chain from the prime conductor, till it comes within a quarter of an inch right over the wire of the bottle ; place your knuckle on the glaſs ſtand, at juſt the ſame diſtance from the coating of the bottle, as the bullet is from its wire. Now let the globe be turned, and you ſee a ſpark ſtrike from the bullet to the wire of the bottle, and the ſame inſtant you ſee and feel an exactly equal ſpark ſtriking from the coating on your knuckle, and ſo on, ſpark for ſpark. This looks as if the whole received by the bottle was again diſcharged from it. And yet the bottle by this means is charged !* And therefore the fire that thus leaves the bottle, though the ſame in quantity, cannot be the very ſame fire that entered at the wire, for if it were, the bottle would remain uncharged.

30. If the fire that ſo leaves the bottle be not the ſame that is thrown in through the wire, it muſt be fire that ſub-ſiſted in the bottle, (that is, in the glaſs of the bottle) be-fore the operation began.

31. If ſo, there muſt be a great quantity in glaſs, be-cauſe a great quantity is thus diſcharged, even from very thin glaſs.

* See Sect. 10, of *Farther Experiments, &c.*

32. That

32. That this electrical fluid or fire is ſtrongly attracted by glaſs, we know from the quickneſs and violence with which it is reſumed by the part that had been deprived of it, when there is an opportunity. And by this, that we cannot from a maſs of glaſs, draw a quantity of electric fire, or electrify the whole maſs *minus,* as we can a maſs of metal. We cannot leſſen or increaſe its whole quantity, for the quantity it has it holds ; and it has as much as it can hold. Its pores are filled with it as full as the mutual repellency of the particles will admit ; and what is already in, refuſes, or ſtrongly repels, any additional quantity. Nor have we any way of moving the electrical fluid in glaſs, but one ; that is, by covering part of the two ſurfaces of thin glaſs with non-electrics, and then throwing an additional quantity of this fluid on one ſurface, which ſpreading in the non-electric, and being bound by it to that ſurface, acts by its repelling force on the particles of the electrical-fluid contained in the other ſurface, and drives them out of the glaſs into the non-electric on that ſide, from whence they are diſcharged, and then thoſe added on the charged ſide can enter. But when this is done, there is no more in the glaſs, nor leſs than before, juſt as much having left it on one ſide as it received on the other.

33. I feel a want of terms here, and doubt much whether I ſhall be able to make this part intelligible. By the word *ſurface,* in this caſe, I do not mean mere length and breadth without thickneſs ; but when I ſpeak of the upper or under ſurface of a piece of glaſs, the outer or in-

ner

ner furface of the vial, I mean length, breadth, and half the thicknefs, and beg the favour of being fo underftood. Now, I fuppofe, that glafs in its firft principles, and in the furnace, has no more of this electrical fluid than other common matter : That when it is blown, as it cools, and the particles of common fire leave it, its pores become a vacuum : That the component parts of glafs are **extremely** fmall and fine, I guefs from its never fhowing a rough face when it breaks, but always a polifh ; and from the fmallnefs of its particles I fuppofe the pores between them muft be exceeding fmall, which is the reafon that aquafortis, nor any other menftruum we have, can enter to feparate them and diffolve the fubftance ; nor is any fluid we know of, fine enough to enter, except common fire, and the electric fluid. Now the departing fire leaving a vacuum, as aforefaid, between thefe pores, which air nor water are fine enough to enter and fill, the electric fluid, (which is every where ready in what we call the non-electrics, and in the non-electric mixtures that are in the air) is attracted in ; yet does not become fixed with the fubftance of the glafs, but fubfifts there as water in a porous ftone, retained only by the attraction of the fixed parts, itfelf ftill loofe and a fluid. But I fuppofe farther, that in the cooling of the glafs, its texture becomes clofeft in the middle, and forms a kind of partition, in which the pores are fo narrow, that the particles of the electrical fluid, which enter both furfaces at the fame time, cannot go through, or pafs and repafs from one furface to the

other,

other, and so mix together ; yet, though the particles of electric fluid, imbibed by each surface, cannot themselves pass through to those of the other, their repellency can, and by this means they act on one another. The particles of the electric fluid have a mutual repellency, but by the power of attraction in the glass they are condensed or forced nearer to each other. When the glass has received, and, by its attraction, forced closer together so much of this electric fluid, as that the power of attracting and condensing in the one, is equal to the power of expansion in the other, it can imbibe no more, and that remains its constant whole quantity ; but each surface would receive more, if the repellency of what is in the opposite surface did not resist its entrance. The quantities of this fluid in each surface being equal, their repelling action on each other is equal ; and therefore those of one surface cannot drive out those of the other ; but, if a greater quantity is forced into one surface than the glass would naturally draw in, this increases the repelling power on that side, and overpowering the attraction on the other, drives out part of the fluid that had been imbibed by that surface, if there be any non-electric ready to receive it : such there is in all cases where glass is electrified to give a shock. The surface that has been thus emptied by having its electrical fluid driven out, resumes again an equal quantity with violence, as soon as the glass has an opportunity to discharge that over quantity more than it could retain by attraction in its other surface, by the additional repellency of which

the

the vacuum had been occafioned. For experiments favouring (if I may not fay confirming) this hypothefis, I muft, to avoid repetition, beg leave to refer you back to what is faid of the electrical phial in my former papers.

34. Let us now fee how it will account for feveral other appearances.—Glafs, a body extremely elaftic (and perhaps its elafticity may be owing in fome degree to the fubfifting of fo great a quantity of this repelling fluid in its pores) muft, when rubbed, have its rubbed furface fomewhat ftretched, or its folid parts drawn a little farther afunder, fo that the vacancies in which the electrical fluid refides, become larger, affording room for more of that fluid, which is immediately attracted into it from the cufhion or hand rubbing, they being fupplied from the common ftock.   But the inftant the parts of the glafs, fo opened and filled, have paffed the friction, they clofe again, and force the additional quantity out upon the furface, where it muft reft till that part comes round to the cufhion again, unlefs fome non-electric (as the prime-conductor) firft prefents to receive it *.   But if the infide of the globe be lined with a non-electric, the additional repellency of the electrical fluid, thus collected

* In the dark the electric fluid may be feen on the cufhion in two femi-circles or half-moons, one on the fore part, the other on the back part of the cufhion, juft where the globe and cufhion feparate.   In the fore crefcent the fire is paffing out of the cufhion into the glafs; in the other it is leaving the glafs, and returning into the back part of the cufhion. When the prime conductor is apply'd to take it off the glafs, the back crefcent difappears.

by

by friction on the rubb'd part of the globe's outer surface, drives an equal quantity out of the inner surface into
that non-electric lining, which receiving it, and carrying
it away from the rubb'd part into the common mass,
through the axis of the globe, and frame of the machine, the new collected electrical fluid can enter and remain in the outer surface, and none of it (or a very little)
will be received by the prime conductor.   As this charg'd
part of the globe comes round to the cushion again, the
outer surface delivers its overplus fire into the cushion, the
opposite inner surface receiving at the same time an equal
quantity from the floor.   Every electrician knows that a
globe wet within will afford little or no fire, but the reason has not before been attempted to be given, that I
know of.

34. So if a tube lined with a * non-electric, be rubb'd,
little or no fire is obtained from it.   What is collected
from the hand in the downward rubbing stroke, entering
the pores of the glass, and driving an equal quantity out
of the inner surface into the non-electric lining : and the
hand in passing up to take a second stroke, takes out
again what had been thrown into the outer surface, and
then the inner surface receives back again what it had
given to the non-electric lining.   Thus the particles of
electrical fluid belonging to the inside surface go in and
out of their pores every stroke given to the tube.   Put a

* Gilt Paper, with the gilt face next the glass, does well.

wire

wire into the tube, the inward end in contact with the non-electric lining, so it will represent the *Leyden* bottle. Let a second person touch the wire while you rub, and the fire driven out of the inward surface when you give the stroke, will pass through him into the common mass, and return through him when the inner surface resumes its quantity, and therefore this new kind of *Leyden* bottle cannot be so charged. But thus it may: after every stroke, before you pass your hand up to make another, let the second person apply his finger to the wire, take the spark, and then withdraw his finger; and so on till he has drawn a number of sparks; thus will the inner surface be exhausted, and the outer surface charged; then wrap a sheet of gilt paper close round the outer surface, and grasping it in your hand you may receive a shock by applying the finger of the other hand to the wire: for now the vacant pores in the inner surface resume their quantity, and the overcharg'd pores in the outer surface discharge that overplus; the equilibrium being restored through your body, which could not be restored through the glass *. If the tube be exhausted of air, a non-electric lining, in contact with the wire, is not necessary; for *in vacuo*, the electrical fire will fly freely from the inner surface, without a non-electric conductor: but air resists in motion; for being itself an electric *per*

* See *Farther Experiments*, Sect. 15.

se,

ſe, it does not attract it, having already its quantity. So the air never draws off an electric atmoſphere from any body, but in proportion to the non-electrics mix'd with it : it rather keeps ſuch an atmoſphere confin'd, which from the mutual repulſion of its particles, tends to diſſipation, and would immediately diſſipate *in vacuo.*—And thus the experiment of the feather incloſed in a glaſs veſſel hermetically ſealed, but moving on the approach of the rubbed tube, is explained : When an additional quantity of the electrical fluid is applied to the ſide of the veſſel by the atmoſphere of the tube, a quantity is repelled and driven out of the inner ſurface of that ſide into the veſſel, and there affects the feather, returning again into its pores, when the tube with its atmoſphere is withdrawn; not that the particles of that atmoſphere did themſelves paſs through the glaſs to the feather.—And every other appearance I have yet ſeen, in which glaſs and electricity are concerned, are, I think, explained with equal eaſe by the ſame hypotheſis. Yet, perhaps, it may not be a true one, and I ſhall be obliged to him that affords me a better.

35. Thus I take the difference between non electrics, and glaſs, an electric *per ſe*, to conſiſt in theſe two particulars. 1ſt, That a non-electric eaſily ſuffers a change in the quantity of the electric fluid it contains. You may leſſen its whole quantity, by drawing out a part, which the whole body will again reſume; but of glaſs you can only leſſen the quantity contained in one of its

M

surfaces ; and not that, but by supplying an equal quantity at the same time to the other surface; so that the whole glass may always have the same quantity in the two surfaces, their two different quantities being added together. And this can only be done in glass that is thin ; beyond a certain thickness we have yet no power that can make this change. And, 2dly, that the electric fire freely removes from place to place, in and through the substance of a non-electric, but not so through the substance of glass. If you offer a quantity to one end of a long rod of metal, it receives it, and when it enters, every particle that was before in the rod, pushes its neighbour quite to the further end, where the overplus is discharged ; and this instantaneously where the rod is part of the circle in the experiment of the shock. But glass, from the smallness of its pores, or stronger attraction of what it contains, refuses to admit so free a motion ; a glass rod will not conduct a shock, nor will the thinnest glass suffer any particle entering one of its surfaces to pass through to the other.

36. Hence we see the impossibility of success in the experiments proposed, to draw out the effluvial virtues of a non-electric, as cinnamon for instance, and mixing them with the electric fluid, to convey them with that into the body, by including it in the globe, and then applying friction, &c. For though the effluvia of cinnamon, and the electric fluid should mix within the globe, they would never come out together through the pores of the

glass,

glafs, and fo go to the prime conductor ; for the electric fluid itfelf cannot come through ; and the prime
conductor is always fupply'd from the cufhion, and that
from the floor. And befides, when the globe is filled
with cinnamon, or other non-electric, no electric fluid
can be obtained from its outer furface, for the reafon before-mentioned. I have tried another way, which I
thought more likely to obtain a mixture of the electric
and other effluvia together, if fuch a mixture had been
poffible. I placed a glafs plate under my cufhion, to cut
off the communication between the cufhion and floor ;
then brought a fmall chain from the cufhion into a glafs
of oil of turpentine, and carried another chain from the
oil of turpentine to the floor, taking care that the chain
from the cufhion to the glafs, touch'd no part of the frame
of the machine. Another chain was fixed to the prime
conductor, and held in the hand of a perfon to be electrifed. The ends of the two chains in the glafs were
near an inch diftant from each other, the oil of turpentine between. Now the globe being turned, could draw
no fire from the floor through the machine, the communication that way being cut off by the thick glafs plate
under the cufhion : it muft then draw it through the
chains whofe ends were dipped in the oil of turpentine.
And as the oil of turpentine, being an electric *per fe,*
would not conduct, what came up from the floor was
obliged to jump from the end of one chain to the end of
the other, through the fubftance of that oil, which we

could

could fee in large fparks, and fo it had a fair opportunity
of feizing fome of the fineft particles of the oil in its paffage, and carrying them off with it : but no fuch effect
followed, nor could I perceive the leaft difference in the
fmell of the electric effluvia thus collected, from what
it has when collected otherwife, nor does it otherwife
affect the body of a perfon electrifed. I likewife put
into a phial, inftead of water, a ftrong purgative liquid,
and then charged the phial, and took repeated fhocks
from it in which cafe every particle of the electrical
fluid muft, before it went through my body, have firft
gone through the liquid when the phial is charging, and
returned through it when difcharging, yet no other effect
followed than if it had been charged with water. I have
alfo fmelt the electric fire when drawn thro' gold, filver,
copper, lead, iron, wood, and the human body, and
could perceive no difference; the odour is always the
fame where the fpark does not burn what it ftrikes ; and
therefore I imagine it does not take that fmell from any
quality of the bodies it paffes through. And indeed, as
that fmell fo readily leaves the electric matter, and adheres to the knuckle receiving the fparks, and to other
things ; I fufpect that it never was connected with it,
but arifes inftantaneoufly from fomething in the air acted
upon by it. For if it was fine enough to come with the
electric fluid through the body of one perfon, why fhould
it ftop on the fkin of another ?

But

But I fhall never have done, if I tell you all my con-jectures, thoughts, and imaginations on the nature and operations of this electric fluid, and relate the variety of little experiments we have tried. I have already made this paper too long, for which I muft crave pardon, not having now time to make it fhorter. I fhall only add, that as it has been obferved here that fpirits will fire by the electric fpark in the fummer time, without heating them, when *Fahrenheit*'s thermometer is above 70; fo when colder, if the operator puts a fmall flat bottle of fpirits in his bofom, or a clofe pocket, with the fpoon, fome little time before he ufes them, the heat of his body will communicate warmth more than fufficient for the purpofe.

A D D I-

ADDITIONAL EXPERIMENT.

Proving that the Leyden Bottle *has no more electrical Fire in it when charged, than before ; nor less when discharged : That, in discharging, the Fire does not issue from the Wire and the Coating at the same Time, as some have thought, but that the Coating always receives what is discharged by the Wire, or an equal Quantity ; the outer Surface being always in a negative State of Electricity, when the inner Surface is in a positive State.*

PLACE a thick plate of glass under the rubbing cushion, to cut off the communication of electrical fire from the floor to the cushion ; then, if there be no fine points or hairy threads sticking out from the cushion, or from the parts of the machine opposite to the cushion, (of which you must be careful) you can get but a few sparks from the prime conductor, which are all the cushion will part with.

Hang a phial then on the prime conductor, and it will not charge though you hold it by the coating.—But

Form a communication by a chain from the coating to the cushion, and the phial will charge.

For the globe then draws the electric fire out of the outside surface of the phial, and forces it through the prime conductor and wire of the phial, into the inside surface.

Thus

Thus the bottle is charged with its own fire, no other being to be had while the glafs plate is under the cufhion.

Hang two cork balls by flaxen threads to the prime conductor.; then touch the coating of the bottle, and they will be electrified and recede from each other.

For juft as much fire as you give the coating, fo much is difcharged through the wire upon the prime conductor, whence the cork balls receive an electrical atmofphere. —But,

Take a wire bent in the form of a C, with a ftick of wax fixed to the outfide of the curve, to hold it by; and apply one end of this wire to the coating, and the other at the fame time to the prime conductor, the phial will be difcharged; and if the balls are not electrified before the difcharge, neither will they appear to be fo after the difcharge, for they will not repel each other.

Now if the fire difcharged from the infide furface of the bottle through its wire, remained on the prime conductor, the balls would be electrified, and recede from each other.

If the phial really exploded at both ends, and difcharged fire from both coating and wire, the balls would be *more* electrified, and recede *farther*; for none of the fire can efcape, the wax handle preventing.

But if the fire, with which the infide furface is furcharged, be fo much precifely as is wanted by the outfide furface, it will pafs round through the wire fixed to the wax

handle

handle, reftore the equilibrium in the glafs, and make no alteration in the ftate of the prime conductor.

Accordingly we find, that if the prime conductor be electrified, and the cork balls in a ftate of repellency before the bottle is difcharged, they continue fo afterwards. If not, they are not electrified by that difcharge.

LET-

LETTER VI.

FROM

BENJ. FRANKLIN, *Esq*; of *Philadelphia*,

TO

PETER COLLINSON, *Esq*; F. R. S. at *London*.

SIR,   *July* 27, 1750.

MR *W-tf-n*, I believe, wrote his Obfervations on my laft paper in hafte, without having firft well confidered the Experiments related §. 17. *. which ftill appear to me decifive in the queftion,—*Whether the accumulation of the electrical fire be in the electrified glafs, or in the non-electric matter connected with the glafs?* and to demonftrate that 'tis really in the glafs.

As to the experiment that ingenious Gentleman mentions, and which he thinks conclufive on the other fide, I perfuade myfelf he will change his opinion of it, when he confiders, that as one perfon applying the wire of the charged bottle to warm fpirits, in a fpoon held by another

* See the Paper entitled, *Farther Experiments, &c.*

N   per-

perfon, both ftanding on the floor, will fire the fpirits, and yet fuch firing will not determine whether the accumulation was in the glafs or the non-electric; fo the placing another perfon between them, ftanding on wax, with a bafon in his hand, into which the water from the phial is pour'd, *while he at the inftant of pouring* prefents a finger of his other hand to the fpirits, does not at all alter the cafe; the ftream from the phial, the fide of the bafon, with the arms and body of the perfon on the wax, being all together but as one long wire, reaching from the internal furface of the phial to the fpirits.

June 29, 1751. In Capt. *Waddell's* account of the effects of lightning on his fhip, I could not but take notice of the large comazants (as he calls them) that fettled on the fpintles at the top-maft heads, and burnt like very large torches (before the ftroke). According to my opinion, the electrical fire was then drawing off, as by points, from the cloud; the largenefs of the flame betokening the great quantity of electricity in the cloud: and had there been a good wire communication from the fpintle heads to the fea, that could have conducted more freely than tarred ropes, or mafts of turpentine wood, I imagine there would either have been no ftroke; or, if a ftroke, the wire would have conducted it all into the fea without damage to the fhip.

His compaffes loft the virtue of the load-ftone, or the poles were reverfed; the North point turning to the South. —By Electricity we have (*here* at *Philadelphia*) frequently
given

given polarity to needles, and reverfed it at pleafure. Mr *Wilfon*, at *London*, tried it on too large maffes, and with too fmall force.

A fhock from four large glafs jars, fent through a fine fewing needle, gives it polarity, and it will traverfe when laid on water.—If the needle when ftruck lies Eaft and Weft, the end entered by the electric blaft points North. —If it lies North and South, the end that lay towards the North will continue to point North when placed on water, whether the fire entered at that end, or at the contrary end.

The Polarity given is ftrongeft when the Needle is ftruck lying North and South, weakeft when lying Eaft and Weft; perhaps if the force was ftill greater, the South end, enter'd by the fire, (when the needle lies North and South) might become the North, otherwife it puzzles us to account for the inverting of compaffes by lightning; fince their needles muft always be found in that fituation, and by our little Experiments, whether the blaft entered the North and went out at the South end of the needle, or the contrary, ftill the end that lay to the North fhould continue to point North.

In thefe experiments the ends of the needles are fometimes finely blued like a watch-fpring by the electric flame. —This colour given by the flafh from two jars only, will wipe off, but four jars fix it, and frequently melt the needles. I fend you fome that have had their heads and points melted off by our mimic lightning; and a pin

N 2　　　　　　that

that had its point melted off, and some part of its head
and neck run. Sometimes the surface on the body of the
needle is also run, and appears blister'd when examined by
a magnifying glass: the jars I make use of hold 7 or 8
gallons, and are coated and lined with tin foil; each of
them takes a thousand turns* of a globe nine inches diameter to charge it.

I send you two specimens of tin-foil melted between
glass, by the force of two jars only.

I have not heard that any of your *European* electricians
have ever been able to fire gunpowder by the electric
flame.—We do it here in this manner.—A small cartridge is filled with dry powder, hard rammed, so as to
bruise some of the grains; two pointed wires are then
thrust in, one at each end, the points approaching each
other in the middle of the cartridge till within the distance
of half an inch; then, the cartridge being placed in the
circle, when the four jars are discharged, the electric
flame leaping from the point of one wire to the point of
the other, within the cartridge amongst the powder, *fires
it*, and the explosion of the powder is at the same instant
with the crack of the discharge.

Yours, &c.

B. FRANKLIN.

* The cushion being afterwards covered with a long flap of buckskin,
which might cling to the globe; and care being taken to keep that flap of
a due temperature; between too dry and too moist, we found so much more
of the electric fluid was obtained, as that 150 turns were sufficient. 1753.

L E T-

LETTER VII.

FROM

Benj. Franklin, *Esq*; of *Philadelphia*,

T O

C. C. Efq; at *New-York*.

S I R,   1751.

I Inclofe you anfwers, fuch as my prefent hurry of bufinefs will permit me to make, to the principal queries contained in yours of the 28th inftant, and beg
leave to refer you to the latter piece in the printed collection of my papers, for farther explanation of the difference between what is called *electrics per fe*, and *non
electrics*. When you have had time to read and confider
thefe papers, I will endeavour to make any new experiments you fhall propofe, that you think may afford farther light or fatisfaction to either of us; and fhall be
much obliged to you for fuch remarks, objections, &c. as
may occur to you.—I forget whether I wrote you that I
have melted brafs pins and fteel needles, inverted the poles
of the magnetic needle, given a magnetifm and polarity to

needles

needles that had none, and fired dry gunpowder by the electric fpark.   I have five bottles that contain 8 or 9 gallons each, two of which charg'd, are fufficient for thofe purpofes : but I can charge and difcharge them altogether.   There are no bounds (but what expence and labour give) to the force man may raife and ufe in the electrical way : For bottle may be added to bottle *in infinitum* and all united and difcharged together as one, the force and effect proportioned to their number and fize.   The greateft known effects of common lightning may, I think, without much difficulty, be exceeded in this way, which a few years fince could not have been believed, and even now may feem to many a little extravagant to fuppofe.—So we are got beyond the fkill of *Rabelais*'s devils of two years old, who, he humoroufly fays, had only learnt to thunder and lighten a little round the head of a cabbage.

I am, with fincere refpect,

Your moft obliged humble fervant,

B. FRANKLIN.

Que-

Queries *and* Anſwers *referr'd to in the forego-ing Letter.*

Query. Wherein confiſts the difference between an *elec-tric* and a *non-electric* body ?

Anſwer. The terms electric *per ſe,* and non-electric, were firſt uſed to diſtinguiſh bodies, on a miſtaken ſup-poſition that thoſe called electrics *per ſe,* alone contained electric matter in their ſubſtance, which was capable of being excited by friction, and of being produced or drawn from them, and communicated to thoſe called non-electrics, ſuppoſed to be deſtitute of it : For the glaſs, &c. being rubbed, diſcover'd ſigns of having it, by ſnapping to the finger, attracting, repelling, &c. and could com-municate thoſe ſigns to metals and water.——Afterwards it was found, that rubbing of glaſs would not produce the electric matter, unleſs a communication was preſerved be-tween the rubber and the floor ; and ſubſequent experi-ments proved that the electric matter was really drawn from thoſe bodies that at firſt were thought to have none in them. Then it was doubted whether glaſs and other bodies called *electrics per ſe,* had really any electric matter in them, ſince they apparently afforded none but what they firſt extracted from thoſe which had been called non-electrics. But ſome of my experiments ſhew that glaſs contains it in great quantity, and I now ſuſpect it to be pretty equally diffuſed in all the matter of this terraqueous

globe

globe· If fo, the terms *electric per fe,* and *non electric,* fhould be laid afide as improper : And (the only difference being this, that fome bodies will conduct electric matter, and others will not) the terms *conductor* and *non-conductor* may fupply their place. If any portion of electric matter is applied to a piece of conducting matter, it penetrates and flows through it, or fpreads equally on its furface ; if applied to a piece of non-conducting matter, it will do neither. Perfect conductors of electric matter are only metals and water. Other bodies conducting only as they contain a mixture of thofe ; without more or lefs of which they will not conduct at all *. This (by the way) fhews a new relation between metals and water heretofore unknown.

To illuftrate this by a comparifon, which, however, can only give a faint refemblance. Electric matter paffes through conductors as water paffes through a porous ftone, or fpreads on their furfaces as water fpreads on a wet ftone ; but when applied to non-conductors, it is like water dropt on a greafy ftone, it neither penetrates, paffes through, nor fpreads on the furface, but remains in drops where it falls. See farther on this head in my laft printed piece.

Query. What are the effects of air in electrical experiments ?

Anfwer. All I have hitherto obferved, are thefe. Moift

air

* This propofition is fince found to be too general ; Mr *Wilfon* having difcovered that melted wax and rofin will alfo conduct.

air receives and conducts the electrical matter in proportion to its moiſture, quite dry air not at all : air is therefore to be claſs'd with the non-conductors. Dry air aſ-
ſiſts in confining the electrical atmoſphere to the body it
ſurrounds, and prevents its diſſipating : for in vacuo it
quits eaſily, and points operate ſtronger, *i. e.* they throw
off or attract the electrical matter more freely, and at
greater diſtances ; ſo that air intervening obſtructs its paſſing from body to body, in ſome degree.　A clean electrical phial and wire, containing air inſtead of water, will
not be charged nor give a ſhock, any more than if it was
fill'd with powder of glaſs ; but exhauſted of air it operates
as well as if filled with water.　Yet, an electric atmoſphere
and air do not ſeem to exclude each other, for we breath
freely in ſuch an atmoſphere, and dry air will blow through
it without diſplacing or driving it away.　I queſtion whether the ſtrongeſt dry N. Weſter would diſſipate it.　I
once electrified a large cork ball, at the end of a ſilk
thread three feet long, the other end of which I held in my
fingers, and whirl'd it round, like a ſling, 100 times in the
air, with the ſwifteſt motion I could poſſibly give it, yet
it retained its electric atmoſphere, though it muſt have
paſſed through 800 yards of air, allowing my arm in
giving the motion to add a foot to the ſemi-diameter of
the circle.—By quite dry air, I mean the dryeſt we have :
for perhaps we never have any perfectly free from moiſture.
An electrical atmoſphere raiſed round a thick wire, inſerted in a phial of air, drives out none of the air, nor

O

on withdrawing that atmofphere will any air ruſh in, as I
have found by a very curious experiment, accurately made,
whence we concluded that the air's elaſticity was not affected thereby.

An Experiment towards difcovering more of the Qualities of
the Electric Fluid.

FROM the prime conductor, hang a bullet by a wire-
hook ; under the bullet at half an inch diſtance, place a
bright piece of ſilver to receive the ſparks ; then let the
wheel be turned, and in a few minutes (if the repeated
ſparks continually ſtrike in the ſame ſpot) the ſilver will
receive a blue ſtain, near the colour of a watch ſpring.

A bright piece of iron will alſo be ſpotted, but not with
that colour ; it rather ſeems corroded.

On gold, braſs, or tin, I have not perceived that it
makes any impreſſion.   But the ſpots on the ſilver or iron
will be the ſame, whether the bullet be lead, braſs, gold,
or ſilver.

On a ſilver bullet there will alſo appear a ſmall ſpot, as
well as on the plate below it.

L E T-

LETTER VIII.

FROM

Mr E. KINNERSLEY at *Boſton,*

TO

BENJAMIN FRANKLIN, Eſq; at *Philadelphia.*

S I R,   *Feb.* 3, 1752.

I Have the following Experiments to communicate : I
held in one hand a wire, which was faſtened at the
other end to the handle of a pump, in order to try
whether the ſtroke from the prime conductor, through
my arms, would be any greater than when conveyed only
to the ſurface of the earth, but could diſcover no difference.

I placed the needle of a compaſs on the point of a long
pin, and holding it in the atmoſphere of the prime conductor, at the diſtance of about three inches, found it to
whirl round like the flyers of a jack, with great rapidity.

I ſuſpended with ſilk a cork ball, about the bigneſs of
a pea, and preſented to it, rubbed amber, ſealing wax,
and ſulphur, by each of which it was ſtrongly repelled;

O 2   then

then I tried rubbed glafs and china, and found that each
of thefe would attract it, until it became electrified again,
and then it would be repelled as at firft ; and while thus
repelled by the rubbed glafs or china, either of the others
when rubbed would attract it.   Then I electrified the ball,
with the wire of a charged phial, and prefented to it
rubbed glafs (the ftopper of a decanter) 'and a china tea-
cup, by which it was as ftrongly repelled as by the wire;
but when I prefented either of the other rubbed electrics,
it would be ftrongly attracted, and when I electrified it
by either of thefe, till it became repelled, it would be attracted by the wire of the phial, but be repelled by its
coating.

Thefe experiments furprized me very much, and have
induced me to infer the following paradoxes.

1. If a glafs globe be placed at one end of a prime
conductor, and a fulphur one at the other end, both being equally in good order, and in equal motion, not a
fpark of fire can be obtained from the conductor ; but one
globe will draw out, as faft as the other gives in. •

2. If a phial be fufpended on the conductor, with a
chain from its coating to the table, and only one of the
globes be made ufe of at a time, 20 turns of the wheel,
for inftance, will charge it; after which, fo many turns
of the other wheel will difcharge it; and as many more
will charge it again.

3. The globes being both in motion, each having a feparate conductor, with a phial fufpended on one of them,
                                                    and

and the chain of it faſtened to the other, the phial will become charged ; one globe charging poſitively, the other negatively.

4. The phial being thus charged, hang it in like manner on the other conductor ; ſet both wheels a going again, and the ſame number of turns that charged it before, will now diſcharge it ; and the ſame number repeated, will charge it again.

5. When each globe communicates with the ſame prime conductor, having a chain hanging from it to the table, one of them, when in motion, (but which I can't ſay) will draw fire up through the cuſhion, and diſcharge it through the chain ; the other will draw it up through the chain, and diſcharge it through the cuſhion.

I ſhould be glad if you would ſend to my houſe for my ſulphur globe, and the cuſhion belonging to it, and make the trial ; but muſt caution you not to uſe chalk on the cuſhion, ſome fine powdered ſulphur will do better. If, as I expect, you ſhould find the globes to charge the prime conductor differently, I hope you will be able to diſcover ſome method of determining which it is that charges poſitively.

I am, &c.

E. KINNERSLEY.

LET-

LETTER IX.

FROM

Benjamin Franklin, *Esq;* at *Philadelphia,*

TO

Mr E. Kinnersley, at *Boston.*

S I R,   *March* 2, 1752.

I Thank you for the Experiments communicated. I sent immediately for your brimstone globe, in order to make the trials you desired, but found it wanted centers, which I have not time now to supply ; but the first leisure I will get it fitted for use, try the experiments, and acquaint you with the result.

In the mean time I suspect, that the different attractions and repulsions you observed, proceeded rather from the greater or smaller quantities of the fire you obtained from different bodies, than from its being of a different *kind*, or having a different *direction*.   In haste,

I am, &c.

B. FRANKLIN.

L E T-

L E T T E R X.

F R O M

BENJAMIN FRANKLIN, *Eſq*; of *Philadelphia,*

T O

Mr E. KINNERSLEY, at *Boſton.*

S I R,   *March* 16, 1752.

HAving brought your brimſtone globe to work, I tried one of the experiments you propoſed, and was agreeably ſurpriſed to find that the glaſs globe being at one end of the conductor, and the ſulphur globe at the other end, both globes in motion, no ſpark could be obtained from the conductor, unleſs when one globe turned ſlower, or was not in ſo good order as the other ; and then the ſpark was only in proportion to the difference, ſo that turning equally, or turning that ſloweſt which worked beſt, would again bring the conductor to afford no ſpark.

I found alſo, that the wire of a phial charg'd by the glaſs globe, attracted a cork ball that had touch'd the wire of a phial charged by the brimſtone globe, and *vice verſa,*

ſo

fo that the cork continued to play between the two phials,
juſt as when one phial was charged through the wire, the
other through the coating, by the glaſs globe alone.   And
two phials charged, the one by the brimſtone globe, the
other by the glaſs globe, would be both difcharged by
bringing their wires together, and ſhock the perſon holding the phials.

From theſe experiments one may be certain that your
2d, 3d, and 4th propoſed experiments, would ſucceed exactly as you ſuppoſe, though I have not tried them, wanting time.—I imagine it is the glaſs globe that charges
poſitively, and the ſulphur negatively, for theſe reaſons,
1. Though the ſulphur globe ſeems to work equally well
with the glaſs one, yet it can never occaſion ſo large and
diſtant a ſpark between my knuckle and the conductor
when the ſulphur one is working, as when the glaſs one
is uſed ; which, I ſuppoſe, is occaſioned by this, that bodies of a certain bigneſs cannot ſo eaſily part with a
quantity of electrical fluid they have and hold attracted
within their ſubſtance, as they can receive an additional
quantity *upon* their ſurface by way of atmoſphere.   Therefore ſo much cannot be drawn *out* of the conductor, as
can be thrown *on* it.   2. I obſerve that the ſtream or
bruſh of fire, appearing at the end of a wire, connected
with the conductor, is long, large, and much diverging,
when the glaſs globe is uſed, and makes a ſnapping (or
rattling) noiſe : but when the ſulphur one is uſed, it is
ſhort, ſmall, and makes a hiſſing noiſe ; and juſt the reverſe

verfe of both happens, when you hold the fame wire in your hand, and the globes are worked alternately : the brufh is large, long, diverging and fnapping (or rattling) when the fulphur globe is turn'd ; fhort, fmall, and hiffing when the glafs globe is turn'd.—When the brufh is long, large, and much diverging, the body to which it joins, feems to me to be throwing the fire out ; and when the contrary appears, it feems to be drinking in.   3. I ob-ferve, that when I hold my knuckle before the fulphur globe, while turning, the ftream of fire between my knuckle and the globe, feems to fpread on its furface, as if it flowed from the finger ; on the glafs globe it is other-wife.   4. The cool wind (or what was called fo) that we ufed to feel as coming from an electrified point, is, I think, more fenfible when the glafs globe is ufed, than when the fulphur one.—But thefe are hafty thoughts. As to your fifth paradox, it muft likewife be true, if the globes are alternately worked ; but if worked together, the fire will neither come up nor go down by the chain, be-caufe one globe will drink it as faft as the other pro-duces it.

I fhould be glad to know whether the effects would be contrary if the glafs globe is folid, and the fulphur globe is hollow ; but I have no means at prefent of trying.

In your journeys, your glafs globes meet with accidents, and fulphur ones are heavy and inconvenient. *Query.* Would not a thin plane of brimftone, caft on a board, ferve on occafion as a cufhion, while a globe of leather

P   ftuffed

ftuffed (properly mounted) might receive the fire from the fulphur, and charge the conductor pofitively? Such a globe would be in no danger of breaking*. I think I can conceive how it may be done; but have not time to add more than that I am,

Yours, &c.

B. FRANKLIN.

The preceding LETTERS *having been tranflated into* French, *and printed at* Paris; *the* Abbe Mazeas, *in a Letter to Dr* Stephen Hales, *dated St* Germain, May 20, 1752, *gives the following Account (printed in the* Philofophical Tranfactions) *of the Experiment made at* Marly, *in purfuance of that propofed by Mr* Franklin, *Page 66.*

S I R,

THE *Philadelphian* experiments, that Mr *Collinfon*, a Member of the Royal Society, was fo kind as to communicate to the public, having been univerfally admired in *France*, the King defired to fee them performed. Wherefore the Duke *D'Ayen* offered his Majefty his country-houfe at St *Germain*, where M. *de Lor*, mafter of Experimental Philofophy, fhould put thofe of *Philadelphia* in execution. His Majefty faw them with great fatisfaction, and greatly applauded Meffieurs *Franklin* and *Collinfon*. Thefe applaufes of his Majefty having excited in

Mef-

---

* The difcoveries of the late ingenious Mr *Symmer*, on the pofitive and negative Electricity produced by the mutual friction of white and black filk, &c. afford hints for farther improvements to be made with this view.

Meffieurs *de Buffon, D'Alibard,* and *De Lor,* a defire of verifying the conjectures of Mr *Franklin,* upon the analogy of thunder and electricity, they prepar'd themfelves for making the experiment.

M. *D'Alibard* chofe, for this purpofe, a garden fituated at *Marly,* where he placed upon an electrical body a pointed bar of iron, of 40 feet high. On the tenth of *May,* 20 minutes paft two in the afternoon, a ftormy cloud having paffed over the place where the bar ftood, thofe that were appointed to obferve it, drew near, and attracted from it fparks of fire, perceiving the fame kind of commotions as in the common electrical Experiments.

M. *de Lor* fenfible of the good fuccefs of this experiment refolved to repeat it at his houfe in the *Eftrapade* at *Paris.* He raifed a bar of iron 99 feet high, placed upon a cake of refin, two feet fquare, and three inches thick. On the 18th of *May,* between four and five in the afternoon, a ftormy cloud having paffed over the bar, where it remained half an hour, he drew fparks from the bar, like thofe from the gun barrel, when, in the electrical experiments the globe is only rubbed by the cufhion, and they produced the fame noife, the fame fire, and the fame crackling. They drew the ftrongeft fparks at the diftance of nine lines, while the rain, mingled with a little hail, fell from the cloud, without either thunder or lightning; this cloud being, according to all appearance, only the confequence of a ftorm, which happened elfewhere.

I am, with a profound refpect,
Your moft humble and obedient fervant.

G. MAZEAS.

A Letter *of Mr* W. WATSON, *F.R.S.* to *the* Royal Society, *concerning the electrical Experiments in* England *upon Thunder-Clouds*. Read Dec. 1752. *Tranſ.* Vol. XLVII.

GENTLEMEN,

AFTER the communications, which we have received from ſeveral of our correſpondents in different parts of the continent, acquainting us with the ſucceſs of their experiments laſt ſummer, in endeavouring to extract the electricity from the atmoſphere during a thunder-ſtorm, in conſequence of Mr *Franklin*'s hypotheſis, it may be thought extraordinary, that no accounts have been yet laid before you, of our ſucceſs here from the ſame experiments. That no want of attention, therefore, may be attributed to thoſe here, who have been hitherto converſant in theſe enquiries, I thought proper to appriſe you, that, though ſeveral members of the Royal Society, as well as myſelf, did, upon the firſt advices from *France*, prepare and ſet up the neceſſary apparatus for this purpoſe, we were defeated in our expectations, from the uncommon coolneſs and dampneſs of the air here, during the whole ſummer. We had only at *London* one thunder ſtorm; viz. on *July* 20; and then the thunder was accompanied with rain ; ſo that, by wetting the apparatus, the electricity was diſſipated too ſoon to be perceived upon touching thoſe parts of the apparatus, which ſerved to

conduct

conduct it. This, I fay, in general prevented our verifying Mr *Franklin*'s hypothefis: But our worthy brother
Mr *Canton* was more fortunate. I take the liberty, therefore, of laying before you an extract of a letter, which I
received from that gentleman, dated from *Spital-fquare,*
July 21, 1752.

" I had yefterday, about five in the afternoon, an op-
" portunity of trying Mr *Franklin*'s experiment of extract-
" ing the electrical fire from the clouds; and fucceeded,
" by means of a tin tube, between three and four feet in
" length, fixed to the top of a glafs one, of about eighteen
" inches. To the upper end of the tin tube, which was
" not fo high as a ftack of chimnies on the fame houfe,
" I faftened three needles with fome wire; and to the
" lower end was folder'd a tin cover to keep the rain from
" the glafs tube, which was fet upright in a block of wood.
" I attended this apparatus as foon after the thunder began
" as poffible, but did not find it in the leaft electrified, till
" between the third and fourth clap; when applying my
" knuckle to the edge of the cover, I felt and heard an
" electrical fpark; and approaching it a fecond time, I
" received the fpark at the diftance of about half an inch,
" and faw it diftinctly. This I repeated four or five times
" in the fpace of a minute; but the fparks grew weaker
" and weaker; and in lefs than two minutes the tin tube
" did not appear to be electrifed at all. The rain con-
" tinued during the thunder, but was confiderably abated
" at the time of making the experiment." Thus far Mr
Canton.   Mr

Mr *Wilson* likewife of the Society, to whom we are much obliged for the trouble he has taken in thefe purfuits, had an opportunity of verifying Mr *Franklin*'s hypothefis. He informed me, by a letter from near *Chelmsford* in *Effex*, dated *Auguft* 12, 1752, that; on that day about noon, he perceived feveral electrical fnaps, during, or rather at the end of a thunder ftorm, from no other apparatus than an iron curtain rod, one end of which he put into the neck of a glafs phial, and held this phial in his hand. To the other end of the iron he faftened three needles with fome filk. This phial, fupporting the rod, he held in one hand, and drew fnaps from the rod with a finger of his other. This experiment was not made upon any eminence, but in the garden of a gentleman, at whofe houfe he then was.

Dr *Bevis* obferved, at Mr *Cave*'s at St *John's Gate*, nearly the fame phænomena as Mr *Canton*, of which an account has been already laid before the public.

Trifling as the effects here mentioned are, when compared with thofe which we have received from *Paris* and *Berlin*, they are the only ones, that the laft fummer here has produced; and as they were made by perfons worthy of credit, they tend to eftablifh the authenticity of thofe tranfmitted from our correfpondents.

I flatter myfelf, that this fhort account of thefe matters will not be difagreeable to you; and am,

with the moft profound Refpect,
Your moft obedient humble Servant,
W. WATSON.

L E T T E R XI.

F R O M

Benj. Franklin, *Efq*; of *Philadelphia*.

Oct. 19, 1752.

AS frequent mention is made in public papers from *Europe* of the fuccefs of the *Philadelphia* experiment for drawing the electric fire from clouds by means of pointed rods of iron erected on high buildings, &c. it may be agreeable to the curious to be informed that the fame experiment has fucceeded in *Philadelphia*, though made in a different and more eafy manner, which is as follows :

Make a fmall crofs of two light ftrips of cedar, the arms fo long as to reach to the four corners of a large thin filk handkerchief when extended ; tie the corners of the handkerchief to the extremities of the crofs, fo you have the body of a kite ; which being properly accommodated with a tail, loop, and ftring, will rife in the air, like thofe made of paper ; but this being of filk, is fitter to bear the wet and wind of a thunder-guft without tearing. To the top of the upright ftick of the crofs is to be fixed a very fharp pointed wire, rifing a foot or more

above

above the wood.   To the end of the twine, next the
hand, is to be tied a filk ribbon, and where the filk and
twine join, a key may be faftened.   This kite is to be
raifed when a thunder guft appears to be coming on, and
the perfon who holds the ftring muft ftand within a door
or window, or under fome cover, fo that the filk ribbon
may not be wet ; and care muft be taken that the twine
does not touch the frame of the door or window.   As
foon as any of the thunder clouds come over the kite, the
pointed wire will draw the electric fire from them, and
the kite, with all the twine, will be electrified, and the
loofe filaments of the twine will ftand out every way, and
be attracted by an approaching finger.   And when the rain
has wet the kite and twine, fo that it can conduct the
electric fire freely, you will find it ftream out plentifully
from the key on the approach of your knuckle.   At this
key the phial may be charged ; and from electric fire thus
obtained, fpirits may be kindled, and all the other electric
experiments be performed, which are ufually done by the
help of a rubbed glafs globe or tube, and thereby the famenefs of the electric matter with that of lightening com-
pletely demonftrated.

                                                    B. F.

L E T-

LETTER XII.

FROM

BENJ. FRANKLIN, *Efq*; of *Philadelphia*,

TO

PETER COLLINSON, *Efq*; F.R.S. *London*.

SIR,   *Philadelphia, September* 1753.

IN my former paper on this fubject, wrote firft in 1747, enlarged and fent to *England* in 1749. I confidered the fea as the grand fource of lightning, imagining its luminous appearance to be owing to electric fire, produc'd by friction between the particles of water and thofe of falt. Living far from the fea, I had then no opportunity of making experiments on the fea water, and fo embraced this opinion too haftily.

For in 1750 and 1751, being occafionally on the fea coaft, I found, by experiments, that fea water in a bottle, tho' at firft it would by agitation appear luminous, yet in a few hours it loft that virtue; *hence, and from this*, that I could not by agitating a folution of fea falt in water pro-

Q

duce

duce any light, I firſt began to doubt of my former hypotheſis, and to ſuſpect that the luminous appearance in ſea
water, muſt be owing to ſome other principles.

I then conſidered whether it were not poſſible, that the
particles of air, being electrics *per ſe*, might, in hard gales
of wind, by their friction againſt trees, hills, buildings,
&c. as ſo many minute electric globes, rubbing againſt non-
electric cuſhions, draw the electric fire from the earth, and
that the riſing vapours might receive that fire from the air,
and, by ſuch means, the clouds become electrified.

If this were ſo, I imagined that by forcing a conſtant
violent ſtream of air againſt my prime conductor, by bellows, I ſhould electrify it *negatively*; the rubbing particles of air, drawing from it part of its natural quantity of
the electric fluid. I accordingly made the experiment, but
it did not ſucceed.

In *September* 1752, I erected an iron rod to draw the
lightning down into my houſe, in order to make ſome experiments on it, with two bells to give notice when the
rod ſhould be electrify'd : A contrivance obvious to every
electrician.

I found the bells rang ſometimes when there was no
lightning or thunder, but only a dark cloud over the rod ;
that ſometimes after a flaſh of lightning they would ſuddenly ſtop ; and, at other times, when they had not rang
before, they would, after a flaſh, ſuddenly begin to ring ;
that the electricity was ſometimes very faint, ſo that when
a ſmall

a fmall fpark was obtain'd, another could not be got for
fome time after; at other times the fparks would follow
extremely quick, and once I had a continual ftream from
bell to bell, the fize of a crow quill : Even during the fame
guft there were confiderable variations.

In the winter following I conceived an experiment, to
try whether the clouds were electrify'd *pofitively* or *negatively*; but my pointed rod, with its apparatus, becoming
out of order, I did not refit it till towards the fpring, when
I expected the warm weather would bring on more frequent thunder-clouds.

The experiment was this: To take two phials; charge
one of them with lightning from the iron rod, and give the
other an equal charge by the electric glafs globe, thro' the
prime conductor : When charg'd, to place them on a table
within three or four inches of each other, a fmall cork ball
being fufpended by a fine filk thread from the ceiling, fo as
it might play between the wires.   If both bottles then were
electrifyed *pofitively*, the ball being attracted and repelled
by one, muft be alfo repell'd by the other.   If the one
pofitively, and the other *negatively*; then the ball would be
attracted and repell'd alternately by each, and continue
to play between them as long as any confiderable charge remained.

Being very intent on making this experiment, it was no
fmall mortification to me, that I happened to be abroad
during two of the greateft thunder-ftorms we had early in
the fpring, and tho' I had given orders in my family, that

if

if the bells rang when I was from home, they fhould catch fome of the lightning for me in electrical phials, and they did fo, yet it was moftly diffipated before my return, and in fome of the other gufts, the quantity of lightning I was able to obtain was fo fmall, and the charge fo weak, that I could not fatisfy myfelf: Yet I fometimes faw what heighten'd my fufpicions, and inflamed my curiofity.

At laft, on the 12th of *April* 1753, there being a.fmart guft of fome continuance, I charged one phial pretty well with lightning, and the other equally, as near as I could judge, with electricity from my glaf globe ; and, having placed them properly, I beheld, with great furprize and pleafure, the cork ball play brifkly between them ; and was convinced that one bottle was electrifed *negatively*.

I repeated this experiment feveral times during the guft, and in eight fucceeding gufts, always with the fame fuccefs ; and being of opinion (for reafons I formerly gave in my letter to Mr *Kinnerfly*, fince printed in *London*) that the glafs globe electrifes *pofitively*, I concluded that the clouds are *always* electrifed *negatively*, or have always in them lefs than their natural quantity of the electric fluid.

Yet notwithftanding fo many experiments, it feems I concluded too foon ; for at laft, *June* the 6th, in a guft which continued from five o'clock, P. M. to feven, I met with one cloud that was electrifed pofitively, tho' feveral that pafs'd over my rod before, during the fame guft, were in the negative ftate. This was thus difcovered :

I had

I had another concurring experiment, which I often re-peated, to prove the negative state of the clouds, *viz.* While the bells were ringing, I took the phial charged from the glass globe, and applied its wire to the erected rod, considering, that if the clouds were electrised *positive-ly*, the rod which received its electricity from them, must be so too ; and then the additional *positive* electricity of the phial would make the bells ring faster :—But, if the clouds were in a *negative* state, they must exhaust the e-lectric fluid from my rod, and bring that into the same ne-gative state with themselves, and then the wire of a positive-ly charg'd phial, supplying the rod with what it wanted, (which it was obliged otherwise to draw from the earth by means of the pendulous brass ball playing between the two bells) the ringing would cease till the bottle was discharg'd.

In this manner I quite discharged into the rod several phials that were charged from the glass globe, the electric fluid streaming from the wire to the rod, 'till the wire would receive no spark from the finger ; and during this supply to the rod from the phial, the bells stopt ringing; but by continuing the application of the phial wire to the rod, I exhausted the natural quantity from the inside sur-face of the same phials, or, as I call it, charged them *ne-gatively.*

At length, while I was charging a phial by my glass globe, to repeat this experiment, my bells, of themselves, stopt ringing, and, after some pause, began to ring again. —But now, when I approached the wire of the charg'd

Q 3

phial

phial to the rod, inftead of the ufual ftream that I expected from the wire to the rod, there was no fpark ; not even
when I brought the wire and the rod to touch ; yet the
bells continued ringing vigoroufly, which proved to me,
that the rod was then *pofitively* electrify'd, as well as the
wire of the phial, and equally fo; and, confequently, that
the particular cloud then over the rod, was in the fame pofitive ftate.   This was near the end of the guft.

But this was a fingle experiment, which, however, deftroys my firft too general conclufion, and reduces me to
this : *That the clouds of a thunder-guft are moft commonly in
a negative ftate of electricity, but fometimes in a pofitive
ftate.*

The latter I believe is rare ; for tho' I foon after the
laft experiment, fet out on a journey to *Bofton*, and was
from home moft part of the fummer, which prevented
my making farther trials and obfervations ; yet Mr *Kinnerfley* returning from the iflands juft as I left home, purfued the experiments during my abfence, and informs me
that he always found the clouds in the *negative* ftate.

So that, for the moft part, in thunder-ftrokes, *'tis the
earth that ftrikes into the clouds, and not the clouds that
ftrike into the earth.*

Thofe who are vers'd in electric experiments, will eafily
conceive, that the effects and appearances muft be nearly
the fame in either cafe ; the fame explofion, and the fame
flafh between one cloud and another, and between the

clouds

clouds and mountains, &c. the same rending of trees, walls, &c. which the electric fluid meets with in its passage, and the same fatal shock to animal bodies ; and that pointed rods fix'd on buildings, or masts of ships, and communicating with the earth or sea, must be of the same service in restoring the equilibrium silently between the earth and clouds, or in conducting a flash or stroke, if one should be, so as to save harmless the house or vessel : For points have equal power to throw off, as to draw on the electric fire, and rods will conduct up as well as down.

But tho' the light gained from these experiments makes no alteration in the practice, it makes a considerable one in the theory.   And now we as much need an hypothesis to explain by what means the clouds become negatively, as before to shew how they became positively electrified.

I cannot forbear venturing some few conjectures on this occasion : They are what occur to me at present, and tho' future discoveries should prove them not wholly right, yet they may in the mean time be of some use, by stirring up the curious to make more experiments, and occasion more exact disquisitions.

I conceive then, that this globe of earth and water, with its plants, animals, and buildings, have, diffus'd throughout their substance, a quantity of the electric fluid, just as much as they can contain, which I call the *natural quantity*.

That this natural quantity is not the same in all kinds of common matter under the same dimensions, nor in the

same

same kind of common matter in all circumstances; but a solid foot, for instance, of one kind of common matter, may contain more of the electric fluid than a solid foot of some other kind of common matter; and a pound weight of the same kind of common matter may, when in a rarer state, contain more of the electric fluid than when in a denser state.

For the electric fluid, being attracted by any portion of common matter, the parts of that fluid (which have among themselves a mutual repulsion) are brought so near to each other by the attraction of the common matter that absorbs them, as that their repulsion is equal to the condensing power of attraction in common matter; and then such portion of common matter will absorb no more.

Bodies of different kinds having thus attracted and absorbed what I call their *natural quantity*, *i. e.* just as much of the electric fluid as is suited to their circumstances of density, rarity, and power of attracting, do not then show any signs of electricity among each other.

And if more electric fluid be added to one of these bodies, it does not enter, but spreads on the surface, forming an atmosphere; and then such body shews signs of electricity.

I have in a former paper compar'd common matter to a sponge, and the electric fluid to water: I beg leave once more to make use of the same comparison, to illustrate farther my meaning in this particular.

When

When a fponge is fomewhat condens'd by being fqueez-
ed between the fingers, it will not receive and retain fo
much water as when in its more loofe and open ftate.

If *more* fqueez'd and condens'd, fome of the water will
come out of its inner parts, and flow on the furface.

If the preffure of the fingers be entirely removed, the
fponge will not only refume what was lately forced out,
but attract an additional quantity.

As the fponge in its rarer ftate will *naturally* attract and
abforb *more* water, and in its denfer ftate will *naturally* attract and abforb *lefs* water ; we may call the quantity it attracts and abforbs in either ftate, its *natural quantity*, the
ftate being confidered.

Now what the fponge is to water, the fame is water to
the electric fluid.

When a portion of water is in its common denfe ftate,
it can hold no more electric fluid than it has ; if any be added, it fpreads on the furface.

When the fame portion of water is rarefy'd into vapour,
and forms a cloud, it is then capable of receiving and abforbing a much greater quantity ; there is room for each
particle to have an electric atmofphere.

Thus water, in its rarefy'd ftate, or in the form of a
cloud, will be in a negative ftate of electricity ; it will have
lefs than its *natural quantity* ; that is, lefs than it is naturally
capable of attracting and abforbing in that ftate.

Such a cloud, then, coming fo near the earth as to be
within the ftriking diftance, will receive from the earth a

R   flafh

flash of the electric fluid ; which flash, to supply a great extent of cloud, must sometimes contain a very great quantity of that fluid.

Or such a cloud, passing over woods of tall trees, may from the points and sharp edges of their moist top leaves, receive silently some supply.

A cloud being by any means supply'd from the earth, may strike into other clouds that have not been supply'd, or not so much supply'd ; and those to others, till an equi-librium is produc'd among all the clouds that are within striking distance of each other.

The cloud thus supply'd, having parted with much of what it first receiv'd, may require and receive a fresh supply from the earth, or from some other cloud, which, by the wind, is brought into such a situation as to receive it more readily from the earth.

Hence repeated and continual strokes and flashes till the clouds have all got nearly their natural quantity as clouds, or till they have descended in showers, and are u-nited again with this terraqueous globe, their original.

Thus thunder-clouds are generally in a negative state of electricity compar'd with the earth, agreeable to most of our experiments ; yet as by one experiment we found a cloud electris'd positively, I conjecture that, in that case, such cloud, after having received what was, in its rare state, only its *natural quantity*, became compress'd by the driving winds, or some other means, so that part of what it had absorb'd was forc'd out, and form'd an electric at-

                                                            mosphere

moſphere around it in its denſer ſtate.   Hence it was capable of communicating poſitive electricity to my rod.

To ſhow that a body in different circumſtances of dila-
tation and contraction is capable of receiving and retaining more or leſs of the electric fluid on its ſurface, I
would relate the following experiment.   I placed a clean
wine glaſs on the floor, and on it a ſmall ſilver can.   In
the can I put about three yards of braſs chain ; to one
end of which I faſtened a ſilk thread, which went right
up to the cieling, where it paſſed over a pulley, and came
down again to my hand, that I might at pleaſure draw
the chain up out of the can, extending it till within a
foot of the cieling, and let it gradually ſink into the can
again.——From the cieling, by another thread of fine
raw ſilk, I ſuſpended a ſmall light lock of cotton, ſo as
that when it hung perpendicularly, it came in contact
with the ſide of the can.——Then approaching the wire
of a charged vial to the can, I gave it a ſpark, which
flow'd round in an electric atmoſphere ; and the lock
of cotton was repelled from the ſide of the can to the
diſtance of about nine or ten inches.   The can would
not then receive another ſpark from the wire of the vial ;
but as I gradually drew up the chain, the atmoſphere of
the can diminiſh'd by flowing over the riſing chain, and
the lock of cotton accordingly drew nearer and nearer to
the can ; and then, if I again brought the vial wire near
the can, it would receive another ſpark, and the cotton
fly off again to its firſt diſtance ; and thus, as the chain

R 2   was

was drawn higher, the can would receive more fparks; becaufe the can and extended chain were capable of fupporting a greater atmofphere than the can with the chain gather'd up into its belly.——And that the atmofphere round the can was diminifhed by raifing the chain, and increafed again by lowering it, is not only agreeable to reafon, fince the atmofphere of the chain muft be drawn from that of the can, when it rofe, and returned to it again when it fell; but was alfo evident to the eye, the lock of cotton always approaching the can when the chain was drawn up, and receding when it was let down again.

Thus we fee that increafe of furface makes a body capable of receiving a greater electric atmofphere: But this experiment does not, I own, fully demonftrate my new hypothefis; for the brafs and filver ftill continue in their folid ftate, and are not rarefied into vapour, as the water is in clouds. Perhaps fome future experiments on vapourized water may fet this matter in a clearer light.

One feemingly material objection arifes to the new hypothefis, and it is this. If water, in its rarefied ftate, as a cloud, requires, and will abford more of the electric fluid than when in its denfe ftate as water, why does it not acquire from the earth all it wants at the inftant of its leaving the furface, while it is yet near, and but juft rifing in vapour? To this difficulty I own I cannot at prefent give a folution fatisfactory to myfelf: I thought, how-

however, that I ought to ftate it in its full force, as I have done, and fubmit the whole to examination.

And I would beg leave to recommend it to the curious in this branch of natural philofophy, to repeat with care and accurate obfervation, the experiments I have reported in this and former papers relating to *pofitive* and *negative* electricity, with fuch other relative ones as fhall occur to them, that it may be certainly known whether the electricity communicated by a glafs globe, be *really pofitive*. And alfo I would requeft all who may have an opportunity of obferving the recent effects of lightning on buildings, trees, &c. that they would confider them particularly with a view to difcover the direction. But in thefe examinations, this one thing is always to be underftood, *viz.* that a ftream of the electric fluid paffing thro' wood, brick, metal, &c. while fuch fluid paffes in *small quantity*, the mutually repulfive power of its parts is confined and overcome by the cohefion of the parts of the body it paffes thro' fo as to prevent an explofion ; but when the fluid comes in a quantity too great to be confin'd by fuch cohefion, it explodes, and rends or fufes the body that endeavour'd to confine it. If it be wood, brick, ftone, or the like, the fplinters will flie off on that fide where there is leaft refiftance. And thus, when a hole is ftruck thro' pafteboard by the electrify'd jar, if the furfaces of the pafte-board are not confin'd or comprefs'd, there will be a bur rais'd all round the hole on both fides the pafte-board; but if one fide be confin'd, fo that the bur cannot be rais'd on that
fide,

fide, it will be all rais'd on the other, which way foever the fluid was directed. For the bur round the outfide of the hole, is the effect of the explofion every way from the center of the ftream, and not an effect of the direction.

In every ftroke of lightning, I am of opinion that the ftream of the electric fluid, moving to reftore the equilibrium between the cloud and the earth, does always previoufly find its paffage, and mark out, as I may fay, its own courfe, taking in its way all the conductors it can find, fuch as metals, damp walls, moift wood, &c. and will go confiderably out of a direct courfe, for the fake of the affiftance of good conductors; and that, in this courfe, it is actually moving, tho' filently and imperceptibly, before the explofion, in and among the conductors; which explofion happens only when the conductors cannot difcharge it as faft as they receive it, by reafon of their being incompleat, difunited, too fmall, or not of the beft materials for conducting. Metalline rods, therefore, of fufficient thicknefs, and extending from the higheft part of an edifice to the ground, being of the beft materials and compleat conductors, will, I think, fecure the building from damage, either by reftoring the equilibrium fo faft as to prevent a ftroke, or by conducting it in the fubftance of the rod as far as the rod goes, fo that there fhall be no explofion but what is above its point, between that and the clouds.

If it be afk'd, what thicknefs of a metalline rod may be fuppofed fufficient? In anfwer, I would remark, that five large glafs jars, fuch as I have defcribed in my former papers,

pers, difcharge a very great quantity of electricity, which
neverthelefs will be all conducted round the corner of a
book, by the fine filletting of gold on the cover, it following the gold the fartheft way about, rather than take the
fhorter courfe through the cover, that not being fo good a
conductor.   Now in this line of gold, the metal is fo extremely thin as to be little more than the colour of gold,
and on an octavo book is not in the whole an inch fquare,
and therefore not the 36th part of a grain according to
M. *Reaumur*; yet 'tis fufficient to conduct the charge of
five large jars, and how many more I know not.   Now, I
fuppofe a wire of a quarter an inch diameter to contain about
5000 times as much metal as there is in that gold line, and
if fo, it will conduct the charge of 25,000 fuch glafs jarrs,
which is a quantity, I imagine, far beyond what was e-
ver contain'd in any one ftroke of natural lightning.   But
a rod of half an inch diameter would conduct four times as
much as one of a quarter.

  And with regard to conducting, tho' a certain thicknefs of metal be required to conduct a great quantity of
electricity, and, at the fame time, keep its own fubftance
firm and unfeparated ; and a lefs quantity, as a very fmall
wire for inftance, will be deftroyed by the explofion ; yet
fuch fmall wire will have anfwered the end of conducting
that ftroke, tho' it become incapable of conducting another.   And confidering the extream rapidity with which
the electric fluid moves without exploding, when it has a
free paffage, or compleat metal communication, I fhould
                                                        think

think a vaſt quantity would be conducted in a ſhort time, either to or from a cloud, to reſtore its equilibrium with the earth, by means of a very ſmall wire ; and therefore thick rods ſhould ſeem not ſo neceſſary.—However, as the quantity of lightning diſcharg'd in one ſtroke, cannot well be meaſured, and, in different ſtrokes, is certainly very various, in ſome much greater than others ; and as iron (the beſt metal for the purpoſe, being leaſt apt to fuſe) is cheap, it may be well enough to provide a larger canal to guide that impetuous blaſt, than we imagine neceſſary : For, though one middling wire may be ſufficient, two or three can do no harm.   And time, with careful obſervations well compar'd, will at length point out the proper ſize to greater certainty.

Pointed rods erected on edifices may likewiſe often prevent a ſtroke, in the following manner.   An eye ſo ſituated as to view horizontally the under ſide of a thunder cloud, will ſee it very ragged, with a number of ſeparate fragments, or petty clouds, one under another, the loweſt ſometimes not far from the earth.   Theſe, as ſo many ſtepping-ſtones, aſſiſt in conducting a ſtroke between the cloud and a building.   To repreſent theſe by an experiment, take two or three locks of fine looſe cotton, connect one of them with the prime conductor by a fine thread of two inches, (which may be ſpun out of the ſame lock by the fingers) another to that, and the third to the ſecond, by like threads.—Turn the globe, and you

will

will fee thefe locks extend themfelves towards the table, (as the lower fmall clouds do towards the earth) being attracted by it: But on prefenting a fharp point erect under the loweft, it will fhrink up to the fecond, the fecond to the firft, and all together to the prime conductor, where they will continue as long as the point continues under them. May not, in like manner, the fmall electrifed clouds, whofe equilibrium with the earth is foon reftor'd by the point, rife up to the main body, and by that means occafion fo large a vacancy, as that the grand cloud cannot ftrike in that place?

Thefe thoughts, my dear friend, are many of them crude and hafty ; and if I were merely ambitious of acquiring fome reputation in philofophy, I ought to keep them by me, till corrected and improved by time and farther experience. But fince even fhort hints and imperfect experiments in any new branch of fcience, being communicated, have oftentimes a good effect, in exciting the attention of the ingenious to the fubject, and fo become the occafion of more exact difquifition, and more compleat difcoveries. You are at liberty to communicate this paper to whom you pleafe ; it being of more importance that knowledge fhould increafe, than that your friend fhould be thought an accurate philofopher.

S	LET-

LETTER XIII.

FROM

BENJ. FRANKLIN, *Esq*, at *Philadelphia*,

TO

PETER COLLINSON, *Esq*; F. R. S. at *London*.

S I R,   *April* 18, 1754.

SIN·CE *September* laſt, having been abroad on two long journeys, and otherwiſe much engag'd, I have made but few obſervations on the *poſitive* and *negative* ſtate of electricity in the clouds.   But Mr *Kinnerſley* kept his rod and bells in good order, and has made many.

Once this winter the bells rang a long time, during a fall of ſnow, tho' no thunder was heard, or lightning ſeen. Sometimes the flaſhes and cracks of the electric matter between bell and bell were ſo large and loud as to be heard all over the houſe : but by all his obſervations, the clouds were conſtantly in a negative ſtate, till about ſix weeks ago, when he found them once to change in a few minutes from the negative to the poſitive.   About a fortnight

night after that he made another obfervation of the fame kind ; and laft *Monday* afternoon, the wind blowing hard at S. E. and veering round to N. E. with many thick driving clouds, there were five or fix fucceffive changes from negative to pofitive, and from pofitive to negative, the bells ftopping a minute or two between every change. Befides the methods mentioned in my paper of *September* laft, of difcovering the electrical ftate of the clouds, the following may be us'd. When your bells are ringing, pafs a rubb'd tube by the edge of the bell, connected with your pointed rod : if the cloud is then in a negative ftate, the ringing will ftop ; if in a pofitive ftate, it will continue, and perhaps be quicker. Or, fufpend a very fmall cork-ball by a fine filk thread, fo that it may hang clofe to the edge of the rod-bell : then whenever the bell is electrified, whether pofitively or negatively, the little ball will be repell'd, and continue at fome diftance from the bell. Have ready a round-headed glafs ftopper of a decanter, rub it on your fide till it is electrified, then prefent it to the cork-ball. If the electricity in the ball is pofitive, it will be repell'd from the glafs ftopper as well as from the bell. If negative, it will fly to the ftopper.

LET-

REMARKS

On the Abbe NOLLET's
Letters on ELECTRICITY.

TO

BENJ. FRANKLIN, *Esq*; of *Philadelphia*,

BY

Mr DAVID COLDEN of *New-York*.

S I R,   *Coldenham, in N. York, Dec.* 4, 1753,

IN confidering the Abbe *Nollet*'s letters to Mr *Frank-lin*, I am obliged to pafs by all the experiments which are made with, or in, bottles hermetically fealed, or exhaufted of air; becaufe, not being able to repeat the experiments, I could not fecond any thing which occurs to me thereon, by experimental proof. Wherefore, the firft point wherein I can dare to give my opinion, is in the Abbe's 4th letter, *p.* 66, where he undertakes to prove, that the electric matter paffes from one furface to another through the intire thicknefs of the glafs: He takes Mr *Franklin*'s experiment of the magical picture, and writes thus of it. " When you electrife a pane of glafs coated

" on

" on both fides with metal, it is evident that whatever is
" placed on the fide oppofite to that which receives the
" electricity from the conductor, receives alfo an evident
" electrical virtue." Which Mr *Franklin* fays, is that e-
qual quantity of electric matter, driven out of this fide,
by what is received from the conductor on the other fide;
and which will continue to give an electrical virtue, to
any thing in contact with it, till it is entirely difcharged
of its electrical fire. To which the Abbe thus objects :
" Tell me, fays he, I pray you, how much time is ne-
" ceffary for this pretended difcharge ? I can affure you,
" that after having maintain'd the electrifation for hours,
" this furface, which ought, as it feems to me, to be en-
" tirely difcharged of its electrical matter, confidering ei-
" ther the vaft number of fparks that were drawn from it,
" or the time that this matter had been expofed to the action
" of the expulfive caufe ; this furface, I fay, appeared ra-
" ther better electrifed thereby, and more proper to pro-
" duce all the effects of an actual electric body. *p.* 68."

The Abbe does not tell us what thofe effects were : all
the effects I could never obferve, and thofe that are to
be obferved can eafily be accounted for, by fuppofing that
fide to be entirely deftitute of electric matter. The moft
fenfible effect of a body charged with electricity is, that
when you prefent your finger to it, a fpark will iffue from
it to your finger : Now when a phial, prepared for the
Leyden experiment, is hung to the gun-barrel or prime-

conductor, and you turn the globe in order to charge it; as foon as the electric matter is excited, you can obferve a fpark to iffue from the external furface of the phial to your finger, which, Mr *Franklin* fays, is the natural electric matter of the glafs driven out by that received by the inner furface from the conductor. If it be only drawn out by fparks, a vaft number of them may be drawn; but if you take hold of the external furface with your hand, the phial will foon receive all the electric matter it is capable of, and the outfide will then be entirely deftitute of its electric matter, and no fpark can be drawn from it by the finger: here then is a want of that effect which all bodies, charg'd with electricity, have. Some of the effects of an electric body, which I fuppofe the Abbe has obferved in the exterior furface of a charged phial, are that all light bodies are attracted by it. This is an effect which I have conftantly obferved, but do not think that it proceeds from an attractive quality in the exterior furface of the phial, but in thofe light bodies themfelves, which feem to be attracted by the phial. It is a conftant obfervation, that when one body has a greater charge of electric matter in it than another (that is in proportion to the quantity they will hold) this body will attract that which has lefs: Now, I fuppofe, and it is a part of Mr *Franklin's* fyftem, that all thofe light bodies which appear to be attracted, have more electric matter in them than the external furface of the phial has, wherefore they endeavour to attract the

phial

phial to them, which is too heavy to be moved by the
fmall degree of force they exert, and yet being greater
than their own weight, moves them to the phial. The
following experiment will help the imagination in conceiving this. Sufpend a cork ball, or a feather by a filk
thread, and electrife it; then bring this ball nigh to any
fixed body, and it will appear to be attracted by that bo-,
dy, for it will fly to it: Now, by the confent of electricians, the attractive caufe is in the ball itfelf, and not
in the fixed body to which it flies : This is a fimilar cafe
with the apparent attraction of light bodies, to the external
furface of a charged phial.

The Abbe fays, *p.* 69. " that he can electrife a hundred
men, ftanding on wax, if they hold hands, and if one of
them touch one of thefe furfaces (the exterior) with the
end of his finger" : This I know he can, while the phial is
charging, but after the phial is charged I am as certain he
cannot: That is, hang a phial, prepared for the *Leyden*
experiment, to the conductor, and let a man, ftanding on
the floor, touch the coating with his finger, while the globe
is turn'd, till the electric matter fpews out of the hook of
the phial, or fome part of the conductor, which I take to
be the certaineft fign that the phial has received all the e-
lectric matter it can: after this appears, let the man, who
before ftood on the floor, ftep on a cake of wax, where
he may ftand for hours, and the globe all that time turned, and yet have no appearance of being electrifed. After

ter the electric matter was spewed out as above from the
hook of a phial prepared for the *Lyden* experiment, I
hung another phial, in like manner prepared, to a hook fixed in the coating of the first, and held this other phial in
my hand; now if there was any electric matter transmit-
ted thro' the glass of the first phial, the second one would
certainly receive and collect it; but having kept the phials
in this situation for a considerable time, during which the
globe was continually turned, I could not perceive that the
second phial was in the least charged, for when I touched the hook with my finger, as in the *Leyden* experiment,
I did not feel the least commotion, nor perceive any spark
to issue from the hook.

I likewise made the following experiment. Having
charged two phials (prepared for the *Leyden* experiment)
through their hooks; two persons took each one of these
phials in their hand; one held his phial by the coating, the
other by the hook, which he could do by removing the communication from the bottom before he took hold of the hook.
These persons placed themselves one on each side of me,
while I stood on a cake of wax, and took hold of the hook
of that phial which was held by its coating (upon which
a spark issued, but the phial was not discharged, as I stood
on wax) keeping hold of the hook, I touched the coating of the phial that was held by its hook with my other
hand, upon which there was a large spark to be seen between my finger and the coating, and both phials were

instantly

inftantly difcharged. If the Abbe's opinion be right, that
the exterior furface, communicating with the coating, is
charged, as well as the interior, communicating with the
hook; how can I, who ftand on wax, difcharge both thefe
phials, when it is well known I could not difcharge one
of them fingly?   Nay, fuppofe I have drawn the electric matter from both of them, what becomes of it ?   For
I appear to have no additional quantity in me when the experiment is over, and I have not ftirr'd off the wax : Wherefore this experiment fully convinces me, that the exterior
furface is not charged ; and not only fo, but that it wants
as much electric matter as the inner has of excefs : For by
this fuppofition, which is a part of Mr *Franklin*'s fyftem,
the above experiment is eafily accounted for, as follows :

When I ftand on wax, my body is not capable of receiving
all the electric matter from the hook of one phial, which
it is ready to give ; neither can it give as much to the coating
of the other phial as it is ready to take, when one is only

T   ap-

applied to me : But when both are applied, the coating
takes from one what the hook gives : Thus I receive the
fire from the firſt phial at B, the exterior ſurface of which
is ſupplied from the hand at A : I give the fire to the ſecond phial at C, whoſe interior ſurface is diſcharged by
the hand at D. This diſcharge at D may be made evident
by receiving that fire into the hook of a third phial, which
is done thus : In place of taking the hook of the ſecond
phial in your hand, run the wire of a third phial, prepared
as for the *Leyden* experiment, through it, and hold this third
phial in your hand, the ſecond one hanging to it, by the
ends of the hooks run through each other : When the experiment is performed, this third phial receives the fire at
D, and will be charged.   When this experiment is conſidered I think, it muſt fully prove that the exterior ſurface
of a charged phial wants electric matter, while the inner
ſurface has an exceſs of it.   One thing more, worthy of
notice in this experiment is, that I feel no commotion or
ſhock in my arms, tho' ſo great a quantity of electric matter paſſes through them inſtantaneouſly : I only feel a prickling in the ends of my fingers.   This makes me think the
Abbe has miſtook, when he ſays, that there is no difference
between the ſhock felt in performing the *Leyden* experiment, and the prickling felt on drawing ſimple ſparks, except that of greater to leſs. In the laſt experiment, as much
electric matter went through my arms, as would have given
me a very ſenſible ſhock, had there been an immediate com-

mu-

munication, by my arms, from the hook to the coating of the same phial ; becaufe when it was taken into a third phial, and that phial difcharged fingly thro' my arms, it gave me a fenfible fhock.   If thefe experiments prove that the electric matter does not pafs through the intire thick-nefs of the glafs ; it is a neceffary confequence that it muft always come out where it enter'd.

The next thing I meet with is in the Abbe's fifth letter *p*. 88, where he differs from Mr *Franklin*, who thinks that the whole power of giving a fhock is in the glafs it-felf, and not in the non-electrics in contact with it.   The experiments which Mr *Franklin* gave to prove this opini-on, in his *Experiments and Obfervations on Electricity, Letter* III. *p*. 24. convinced me that he was in the right ; and what the Abbe has afferted in contradiction thereto, has not made me think otherwife.   The Abbe perceiving as I fuppofe, that the experiments, as Mr *Franklin* had perform'd them, muft prove his affertion ; alters them without giving any reafon for it, and makes them in a manner that proves nothing.   Why will he have the phial, into which the water is to be decanted from a charged phial, held in a man's hand ?   If the power of giving a fhock is in the water contain'd in the phial, it fhould re-main there tho' decanted into another phial, fince no non-electric body touch'd it to take that power off.   The phial being placed on wax is no objection, for it cannot take the power from the water, if it had any, but it is a neceffary

T 2   means

means to try the fact ; whereas, that phial's being charged
when held in a man's hand, only proves that water will
conduct the electric matter.   The Abbe owns, *p.* 94, that
he had heard this remarked, but says, Why is not a conductor of electricity an electric subject ?   This is not the
question ; Mr *Franklin* never said that water was not an e-
lectric subject ; he said, that the power of giving a shock
was in the glass, and not in the water ; and this, his experiments, fully prove ; so fully, that it may appear imper-
tinent to offer any more : Yet as I do not know that the
following has been taken notice of by any body before,
my inserting of it in this place may be excused.   It is this :
Hang a phial, prepared for the *Leyden* experiment,
to the conductor, by its hook, and charge it, which
done, remove the communication from the bottom of
the phial.   Now the conductor shews evident signs of being electrised ; for if a thread be tied round it, and its ends
left about two inches long, they will extend themselves
out like a pair of horns ; but if you touch the conductor, a spark will issue from it, and the threads will fall,
nor does the conductor shew the least sign of being electrised after this is done.   I think that by this touch, I
have taken out all the charge of electric matter that was
in the conductor, the hook of the phial, and water or filings of iron contain'd in it ; which is no more than we
see all non-electric bodies will receive ; yet the glass of the
phial retains its power of giving a shock, as any one will
                                                        find

find that pleafes to try. This experiment fully evidences, that the water in the phial contains no more electric matter than it would do in an open bafon, and has not any of that great quantity which produces the fhock, and is only retain'd by the glafs. If after the fpark is drawn from the conductor, you touch the coating of the phial (which all this while is fuppofed to hang in the air, free from any non-electric body) the threads on the conductor will inftantly ftart up, and fhew that the conductor is electrifed. It receives this electrifation from the inner furface of the phial, which, when the outer furface can receive what it wants from the hand applied to it, will give as much as the bodies in contact with it can receive, or, if they be large enough, all that it has of excefs. It is diverting to fee how the threads will rife and fall by touching the coating and conductor of the phial alternately. May it not be that the difference between the charged fide of the glafs, and the outer or emptied fide, being leffen'd by touching the hook or the conductor; the outer fide can receive from the hand which touched it, and by its receiving the inner fide cannot retain fo much ; and for that reafon fo much as it cannot contain electrifes the water, or filings and conductor: For it feems to be a rule, that the one fide muft be emptied in the fame proportion that the other is fill'd : Tho' this from experiment appears evident, yet it is ftill a myftery not to be accounted for.

I am

I am, in many places of the Abbe's book, furprifed to find that experiments have fucceeded fo differently at *Paris* from what they did with Mr *Franklin*, and as I have always obferv'd them to do. The Abbe, in making experiments to find the difference between the two furfaces of a charged glafs, will not have the phial placed on wax: For, fays he, don't you know that being placed on a body originally electric, it quickly lofes its virtue? I cannot imagine what fhould have made the Abbe think fo; it certainly is contradictory to the notions commonly received of electrics per fe; and by experiment I find it entirely otherwife: For having feveral times left a charged phial, for that purpofe, ftanding on wax for hours, I found it to retain as much of its charge as another that ftood at the fame time on a table. I left one ftanding on wax from 10 o'clock at night till 8 next morning, when I found it to retain a fufficient quantity of its charge, to give me a fenfible commotion in my arms, though the room in which the phial ftood had been fwept in that time, which muft have rais'd much duft to facilitate the difcharge of the phial.

I find that a cork ball fufpended between two bottles, the one fully and the other but little charged, will not play between them, but is driven into a fituation that makes a triangle with the hooks of the phials; though the Abbe has afferted the contrary of this, *p.* 101, in order to account for the playing of a cork ball between the

wire

wire thruft into the phial, and one that rifes up from its coating. The phial which is leaft charged muft have more electric matter given to it, in proportion to its bulk, than the cork ball receives from the hook of the full phial.

The Abbe fays, *p.* 103, "that a piece of metal leaf "hung to a filk thread and electrifed, will be repell'd by "the bottom of a charged phial held by its hook in the "air:" This I find conftantly otherwife, it is with me always firft attracted and then repelled : It is neceffary in charging the leaf to be careful, that it does not fly off to fome non-electric body, and fo difcharge itfelf when you think it is charged ; it is difficult to keep it from flying to your own wrift, or to fome part of your body.

The Abbe, *p.* 108, fays, "that it is not impoffible, as "Mr *Franklin* fays it is, to charge a phial while there is a "communication form'd between its coating and its hook." I have always found it impoffible to charge fuch a phial fo as to give it a fhock : Indeed if it hang on the conductor without a communication from it, you may draw a fpark from it as you may from any body that hangs there, but this is very different from being charged in fuch a manner as to give a fhock. The Abbe, in order to account for the little quantity of electric matter that is to be found in the phial, fays, "that it rather follows the metal than the "glafs, and that it is fpewed out into the air from the coating "of the phial" I wonder how it comes not to do fo too,

when

when it fifts through the glafs, and charges the exterior furface, according to the Abbe's fyftem !

The Abbe's objections againft Mr *Franklin*'s two laft experiments, I think, have little weight in them : He feems, indeed, much at a lofs what to fay, wherefore he taxes Mr *Franklin* with having conceal'd a material part of the experiment ; a thing too mean for any gentleman to be charged with, who has not fhewn as great a partiality in relating experiments, as the Abbe has done.

ELEC-

ELECTRICAL EXPERIMENTS,

With an Attempt to account for their

SEVERAL PHÆNOMENA.

Together with

Some Obſervations on *Thunder-Clouds,*

In further Confirmation of Mr FRANKLIN's Obſervations on the poſitive and negative electrical State of the Clouds, by JOHN CANTON, M. A. and F. R. S.

Dec. 6, 1753.

EXPERIMENT I.

FROM the cieling, or any convenient part of a room, let two cork-balls, each about the bigneſs of a ſmall pea, be ſuſpended by linen threads of eight or nine inches in length, ſo as to be in contact with each other. Bring the excited glaſs tube under the balls,

U   and

and they will be feparated by it, when held at the diftance
of three or four feet ; let it be brought nearer, and they
will ftand farther apart , intirely withdraw it, and they,
will immediately come together.   This experiment may
be made with very fmall brafs balls hung by filver wire ;
and will fucceed as well with fealing-wax made electrical,
as with glafs.

EXPERIMENT II.

If two cork-balls be fufpended by dry filk threads, the
excited tube muft be brought within eighteen inches before they will repel each other ; which they will continue
to do, for fome time, after the tube is taken away.

As the balls in the firft experiment are not infulated,
they cannot properly be faid to be electrified : but when
they hang within the atmofphere of the excited tube, they
may attract and condenfe the electrical fluid round about
them, and be feparated by the repulfion of its particles.
It is conjectur'd alfo, that the balls at this time contain
lefs than their common fhare of the electrical fluid, on
account of the repelling power of that which furrounds
them ; tho' fome, perhaps, is continually entering and paffing thro' the threads.   And if that be the cafe, the reafon is plain why the balls hung by filk, in the fecond experiment, muft be in a much more denfe part of the atmofphere of the tube, before they will repel each other.
At the approach of an excited ftick of wax to the balls,
in the firft experiment, the electrical fire is fuppofed to
come

come through the threads into the balls, and be condenfed there, in its paffage towards the wax ; for, according to Mr *Franklin*, excited glafs *emits* the electrical fluid, but excited wax *receives* it.

EXPERIMENT III.

Let a tin tube, of four or five feet in length, and about two inches in diameter, be infulated by filk ; and from one end of it let the cork-balls be fufpended by linen threads. Electrify it, by bringing the excited glafs tube near the other end, fo as that the balls may ftand an inch and an half, or two inches, a-part: Then, at the approach of the excited tube, they will, by degrees, lofe their repelling power, and come into contact; and as the tube is brought ftill nearer, they will feparate again to as great a diftance as before : In the return of the tube they will approach each other till they touch, and then repel as at firft.   If the tin tube be electrified by wax, or the wire of a charg'd phial, the balls will be affected in the fame manner at the approach of excited wax, or the wire of the phial.

EXPERIMENT IV.

Electrify the cork-balls as in the laft experiment by glafs, and at the approach of an excited ftick of wax their re- pulfion will be increafed.   The effect will be the fame, if the excited glafs be brought towards them, when they have been electrified by wax.

The bringing the excited glafs to the end, or edge of

the

the tin-tube, in the third experiment, is fuppos'd to electrify it pofitively, or to add to the electrical fire it before
contained ; and therefore fome will be running off through
the balls, and they will repel each other.   But at the approach of excited glafs, which likewife *emits* the electrical
fluid, the difcharge of it from the balls will be diminifh'd ;
or part will be driven back, by a force acting in a contrary
direction ; and they will come nearer together.   If the
tube be held at fuch a diftance from the balls, that the
excefs of the denfity of the fluid round about them, above
the common quantity in air, be equal to the excefs of the
denfity of that within them, above the common quantity
contain'd in cork ; their repulfion will be quite deftroy'd.
But if the tube be brought nearer ; the fluid wtthout, being more denfe than that within the balls, it will be attracted by them, and they will recede from each other
again.

When the apparatus has loft part of its natural fhare
of this fluid, by the approach of excited wax to one end
of it, or is electrified negatively ; the electrical fire is attracted and imbib'd by the balls to fupply the deficiency ;
and that more plentifully at the approach of excited glafs ;
or a body pofitively electrified, than before ; whence the
diftance between the balls will be increafed, as the fluid
furrounding them is augmented.   And in general, whether by the approach or recefs of any body ; if the difference between the denfity of the internal and external fluid
                                                         be

be increafed, or diminifhed ; the repulfion of the balls will be increafed, or diminifhed, accordingly.

EXPERIMENT V.

When the infulated tin tube is not electrified, bring the excited glafs tube towards the middle of it, fo as to be nearly at right angles with it, and the balls at the end will repel each other ; and the more fo, as the excited tube is brought nearer. When it has been held a few feconds, at the diftance of about fix inches, withdraw it, and the balls will approach each other till they touch ; and then feparating again, as the tube is moved farther off, will continue to repel when it is taken quite away. And this repulfion between the balls will be increafed by the approach of excited glafs, but diminifhed by excited wax ; juft as if the apparatus had been electrified by wax, after the manner defcribed in the third experiment.

EXPERIMENT VI.

Infulate two tin tubes, diftinguifhed by *A* and *B*, fo as to be in a line with each other, and about half an inch apart ; and at the remote end of each, let a pair of cork balls be fufpended. Towards the middle of *A*, bring the excited glafs tube, and holding it a fhort time, at the diftance of a few inches, each pair of balls will be obferved to feparate : withdraw the tube, and the balls of *A* will come together, and then repel each other again ; but thofe of *B* will hardly be affected. By the approach of the excited

cited glafs tube, held under the balls of *A*, their repulfion will be increafed : but if the tube be brought, in the fame manner, towards the balls of *B*, their repulfion will be diminifhed.

In the fifth experiment, the common ftock of electrical matter in the tin tube, is fuppofed to be attenuated about the middle, and to be condenfed at the ends, by the repelling power of the atmofphere of the excited glafs tube, when held near it. And perhaps the tin tube may lofe fome of its natural quantity of the electrical fluid, before it receives any from the glafs; as that fluid will more readily run off from the ends and edges of it, than enter at the middle : and accordingly, when the glafs tube is withdrawn, and the fluid is again equally diffufed through the apparatus, it is found to be electrified negatively : For excited glafs brought under the balls will increafe their repulfion.

In the fixth experiment, part of the fluid driven out of one tin tube enters the other; which is found to be electrified pofitively, by the decreafing of the repulfion of its balls, at the approach of excited glafs.

E X P E R I M E N T   VII.

Let the tin tube, with a pair of balls at one end, be placed three feet at leaft from any part of the room, and the air render'd very dry by means of a fire : electrify the apparatus to a confiderable degree; then touch the tin tube with a finger, or any other conductor, and the balls

will

will, notwithstanding, continue to repel each other; tho' not at so great a great a distance as before.

The air surrounding the apparatus to the distance of two or three feet, is supposed to contain more or less of the electrical fire, than its common share, as the tin tube is electrified positively, or negatively; and when very dry, may not part with its overplus, or have its deficiency supplied so suddenly, as the tin; but may continue to be electrified, after that has been touch'd for a considerable time.

EXPERIMENT VIII.

Having made the Torricellian vacuum about five feet long, after the manner described in the *Philosophical Transactions*, Vol. xlvii. p. 370. if the excited tube be brought within a small distance of it, a light will be seen through more than half its length; which soon vanishes, if the tube be not brought nearer; but will appear again, as that is moved farther off. This may be repeated several times, without exciting the tube afresh.

This experiment may be consider'd as a kind of ocular demonstration of the truth of Mr *Franklin*'s hypothesis; that when the electrical fluid is condensed on one side of thin glass, it will be repelled from the other, if it meets with no resistance. According to which, at the approach of the excited tube, the fire is supposed to be repelled from the inside of the glass surrounding the vacuum, and to be

carried

carried off through the columns of mercury ; but, as the tube is withdrawn, the fire is fuppofed to return.

EXPERIMENT IX.

Let an excited ftick of wax, of two feet and an half in length, and about an inch in diameter, be held near its middle. Excite the glafs tube, and draw it over one half of it ; then, turning it a little about its axis, let the tube be excited again, and drawn over the fame half; and let this operation be repeated feveral times : then will that half deftroy the repelling power of balls electrified by glafs, and the other half will increafe it.

By this experiment it appears, that wax alfo may be e-lectrified pofitively and negatively. And it is probable, that all bodies whatfoever may have the quantity they contain of the electrical fluid, increafed, or diminifhed, The clouds, I have obferved, by a great number of expe-riments, to be fome in a pofitive, and others in a negative ftate of electricity. For the cork balls, electrified by them, will fometimes clofe at the approach of excited glafs ; and at other times be feparated to a greater diftance. And this change I have known to happen five or fix times in lefs than half an hour ; the balls coming together each time and remaining in contact a few feconds, before they repel each other again. It may likewife eafily be difcover'd, by a charged phial, whether the electrical fire be drawn out of the apparatus by a negative cloud, or forced into it by

a pofi-

a pofitive one : and by which foever it be electrified, fhould that cloud either part with its overplus, or have its deficiency fupplied fuddenly, the apparatus will lofe its electricity : which is frequently obferved to be the cafe, immediately after a flafh of lightning. Yet when the air is very dry, the apparatus will continue to be electrifed for ten minutes, or a quarter of an hour, after the clouds have paffed the zenith ; and fometimes till they appear more than half-way towards the horizon. Rain, efpecially when the drops are large, generally brings down the electrical fire : and hail, in fummer, I believe never fails. When the apparatus was laft electrified, it was by the fall of thawing fnow, which happened fo lately, as on the 12th of *November* ; that being the twenty-fixth day, and fixty-firft time, it has been electrified, fince it was firft fet up ; which was about the middle of *May*. And as *Fahrenheit*'s thermometer was but feven degrees above freezing, it is fuppofed the winter will not intirely put a ftop to obfervations of this fort. At *London*, no more than two thunderftorms have happened during the whole fummer ; and the apparatus was fometimes fo ftrongly electrified in one of them, that the bells, which have been frequently rung by the clouds, fo loud as to be heard in every room of the houfe (the doors being open) were filenced by the almoft conftant ftream of denfe electrical fire, between each bell and the brafs ball, which would not fuffer it to ftrike.

I fhall conclude this paper, already too long, with the following queries :

X

1. May not air, fuddenly rarefied, give electrical fire to, and air fuddenly condenfed, receive, electrical fire from, clouds and vapours paffing through it?

2. Is not the *aurora borealis*, the flafhing of electrical fire from pofitive, towards negative clouds at a great diftance, through the upper part of the atmofphere, where the refiftance is leaft?

A P-

APPENDIX.

AS Mr *Franklin*, in a former letter to Mr *Collin-son*, mentioned his intending to try the power of a very ſtrong electrical ſhock upon a turkey, that gentleman accordingly has been ſo very obliging as to ſend an account of it, which is to the following purpoſe.

He made firſt ſeveral experiments on fowls and found, that two large thin glaſs jars gilt, holding each about ſix gallons, were ſufficient, when fully charged, to kill common hens outright; but the turkeys, though thrown into violent convulſions, and then lying as dead for ſome minutes, would recover in leſs than a quarter of an hour. However, having added three other ſuch to the former two, though not fully charged, he killed a turkey of about ten pounds weight, and believes that they would have killed a much larger. He conceited, as himſelf ſays, that the birds kill d in this manner eat uncommonly tender.

In making theſe experiments, he found, that a man could, without great detriment, bear a much greater ſhock than he had imagined: for he inadvertently received the ſtroke of two of theſe jars through his arms and body, when they were very near fully charged. It ſeemed to him an univerſal blow throughout the body from head to foot,

and

and was followed by a violent quick trembling in the trunk, which went off gradually in a few seconds. It was some minutes before he could recollect his thoughts, so as to know what was the matter; for he did not see the flash, tho' his eye was on the spot of the prime-conductor, from whence it struck the back of his hand; nor did he hear the crack, though the by standers said it was a loud one; nor did he particularly feel the stroke on his hand, tho' he afterwards found it had raised a swelling there, of the bigness of half a pistol-bullet. His arms and the back of the neck felt somewhat numbed the remainder of the evening, and his breast was sore for a week after, as if it had been bruised. From this experiment may be seen the danger, even under the greatest caution, to the operator, when making these experiments with large jars; for it is not to be doubted, but several of these fully charged would as certainly, by increasing them, in proportion to the size, kill a man, as they before did a turkey.

N. B. The original of this letter, which was read at the Royal Society, has been mislaid.

ELECTRICAL *and other* PHILOSOPHICAL PAPERS *and* LETTERS.

ELECTRICAL EXPERIMENTS *made in Purfuance of thofe made by Mr* Canton, *dated* December 6, 1753; *with Explanations, by Mr* Benjamin Franklin.

Philadelphia, March 14, 1755.

PRINCIPLES.

Read at the Royal Society, *Dec.* 18, 1755.

I. ELECTRIC atmofpheres, that flow round non-electric bodies, being brought near each other, do not readily mix and unite into one atmofphere, but remain feparate, and repel each other.

This is plainly feen in fufpended cork balls, and other bodies electrified.

II. An electric atmofphere not only repels another electric atmofphere, but will alfo repel the electric matter contained in the fubftance of a body approaching it; and without joining or mixing with it, force it to other parts of the body that contained it.

Y

This

This is fhewn by fome of the following experiments.

III. Bodies electrified negatively, or deprived of their natural quantity of Electricity, repel each other, (or at leaft appear to do fo, by a mutual receding) as well as thofe electrified pofitively, or which have electric atmo-fpheres.

This is fhewn by applying the negatively charged wire of a phial to two cork balls, fufpended by filk threads, and by many other experiments.

PREPARATION.

Fix a taffel of fifteen or twenty threads, three inches long at one end, of a tin prime conductor, (mine is about five feet long, and four inches diameter) fupported by filk lines.

Let the threads be a little damp, but not wet.

EXPERIMENT I.

Pafs an excited glafs Tube near the other end of the prime conductor, fo as to give it fome fparks, and the threads will diverge.

Becaufe each thread, as well as the prime-conductor, has acquired an electric atmofphere, which repels and is re-pelled by the atmofpheres of the other threads: if thofe fe-veral atmofpheres would readily mix, the threads might unite, and hang in the middle of one atmofphere, common to them all.

Rub

Rub the tube afrefh, and approach the prime-conductor there-
with, crofsways, near that end, but not nigh enough to
give fparks; and the threads will diverge a little more.

Becaufe the atmofphere of the prime-conductor is preffed by the atmofphere of the excited tube, and driven towards the end where the threads are, by which each thread
acquires more atmofphere.

Withdraw the tube, and they will clofe as much.

They clofe as much, and no more ; becaufe the atmofphere of the glafs tube not having mixed with the atmofphere of the prime conductor, is withdrawn intire, having
made no addition to, or diminution from it.

Bring the excited tube under the tuft of threads, and they
will clofe a little.

They clofe, becaufe the atmofphere of the glafs tube repels their atmofpheres, and drives part of them back on the
prime conductor.

Withdraw it, and they will diverge as much.

For the portion of atmofphere which they had loft, returns to them again.

EXPERIMENT II.

Excite the glafs tube, and approach the prime conductor with
it, holding it acrofs, near the end oppofite to that on which
the threads hang, at the diftance of five or fix inches.
Keep it there a few feconds, and the threads of the taf-
fels will diverge. Withdraw it, and they will clofe.

They

They diverge, becaufe they have received electric atmofpheres from the electric matter before contained in the fubftance of the prime conductor ; but which is now repelled
and driven away, by the atmofphere of the glafs tube,
from the parts of the prime conductor oppofite and nearest
to that atmofphere, and forced out upon the furface of
the prime conductor at its other end, and upon the threads
hanging thereto. Were it any part of the atmofphere of
the glafs tube that flowed over and along the prime conductor to the threads, and gave them atmofpheres, (as is
the cafe when a fpark is given to the prime conductor from
the glafs tube) fuch part of the tube's atmofphere would
have remained, and the threads continue to diverge ; but
they clofe on withdrawing the tube, becaufe the tube takes
with it *all its own atmofphere*, and the electric matter,
which had been driven out of the fubftance of the prime
conductor, and formed atmofpheres round the threads, is
thereby permitted to return to its place.

*Take a fpark from the prime conductor near the threads,
when they are diverged as before, and they will clofe.*

For by fo doing they take away their atmofpheres, compofed of the electric matter driven out of the fubftance of
the prime conductor, as aforefaid, by the repellency of
the atmofphere of the glafs tube. By taking this fpark
you rob the prime conductor of part of its natural quantity of the electric matter ; which part fo taken is not fupplied by the glafs tube, for when that is afterwards withdrawn

drawn, it takes with it its whole atmofphere, and leaves the prime conductor electrifed negatively, as appears by the next operation.

Then withdraw the tube, and they will open again.

For now the electric matter in the prime conductor, returning to its equilibrium, or equal diffufion, in all parts of its fubftance, and the prime conductor having loft fome of its natural quantity, the threads connected with it lofe part of theirs, and fo are electrifed negatively, and therefore repel each other, by *Pr.* III.

Approach the prime conductor with the tube near the fame place as at firft, and they will clofe again.

Becaufe the part of their natural quantity of electric fluid, which they had loft, is now reftored to them again, by the repulfion of the glafs tube forcing that fluid to them from other parts of the prime conductor ; fo they are now again in their natural ftate.

Withdraw it, and they will open again.

For what had been reftored to them, is now taken from them again, flowing back into the prime conductor, and leaving them once more electrifed negatively.

Bring the excited tube under the threads, and they will diverge more.

Becaufe more of their natural quantity is driven from them into the prime conductor, and thereby their negative Electricity increafed.

E X-

EXPERIMENT III.

The prime conductor not being lectrified, bring the excited tube under the taſſel, and the threads will diverge.

Part of their natural quantity is thereby driven out of them into the prime conductor, and they become negatively electriſed, and therefore repel each other.

Keeping the tube in the ſame place with one hand, attempt to touch the threads with the finger of the other hand, and they will recede from the finger.

Becauſe the finger being plunged into the atmoſphere of the glaſs tube, as well as the threads, part of its natural quantity is driven back through the hand and body, by that atmoſphere, and the finger becomes, as well as the threads, negatively electriſed, and ſo repels, and is repelled by them. To confirm this, hold a ſlender light lock of cotton, two or three inches long, near a prime conductor, that is electrified by a glaſs globe, or tube. You will ſee the cotton ſtretch itſelf out towards the prime conductor. Attempt to touch it with the finger of the other hand, and it will be repelled by the finger. Approach it with a poſitively charged wire of a bottle, and it will fly to the wire. Bring it near a negatively charged wire of a bottle, it will recede from that wire in the ſame manner that it did from the finger ; which demonſtrates the finger to be negatively electriſed, as well as the lock of cotton ſo ſituated.

Extract

Extract of a Letter concerning Electricity, from Mr B. Franklin, *to Monf.* Dalibard, *at* Paris, *inclosed in a Letter to Mr* Peter Collinson, F. R. S.

Philadelphia, June 29, 1755.

Read at the Royal
Society, *Dec.*
18, 1755.

YOU desire my opinion of *Pere Beccaria's Italian* book *. I have read it with much pleasure, and think it one of the best pieces on the subject that I have seen in any language. Yet as to the article of water-spouts, I am not at present of his sentiments; though I must own with you, that he has handled it very ingeniously. Mr *Collinson* has my opinion of whirlwinds and water-spouts at large, written some time since. I know not whether they will be published; if not, I will get them transcribed for your perusal. It does not appear to me that *Pere Beccaria* doubts of the *absolute impermeability of glass* in the sense I meant it; for the instances he gives of holes made through glass by the electric stroke, are such as we have all experienced, and only shew that the electric fluid could not pass without making a hole. In the same manner we say, glass is impermeable to water, and yet a stream from a fire engine will force through the strongest panes of a window. As to the effect of points in

---

* This work is written conformable to Mr *Franklin's* theory, upon artificial and natural Electricity, which compose the two parts of it. It was printed in *Italian* at *Turin,* in 4to. 1753; between the two parts is a letter to the Abbe *Nollet,* in defence of Mr *Franklin's* system. J. B.

drawing

drawing the electric matter from clouds, and thereby fecuring buildings, &c. which, you fay, he feems to doubt, I
muft own I think he only fpeaks modeftly and judicioufly.
I find I have been but partly underftood in that matter.
I have mentioned it in feveral of my letters, and except
once, always in the *alternative, viz.* that pointed rods e-
rected on buildings, and communicating with the moift
earth, would either *prevent* a ftroke, *or*, if not prevented,
would *conduct* it, fo as that the building fhould fuffer no
damage. Yet whenever my opinion is examined in *Europe*, nothing is confidered but the probability of thofe rods
preventing a ftroke or explofion, which is only a *part* of
the ufe I propofed for them; and the other part, their
conducting a ftroke, which they may happen not to prevent, feems to be totally forgotten, though of equal importance and advantage.

I thank you for communicating M. *de Buffon*'s relation
of the effect of lightning at *Dijon*, on the 7th of *June* laft.
In return, give me leave to relate an inftance I lately faw
of the fame kind. Being in the town of *Newbury* in *New-
England*, in *November* laft, I was fhewn the effect of
lightning on their church, which had been ftruck a few
months before. The fteeple was a fquare tower of wood,
reaching feventy feet up from the ground to the place
where the bell hung, over which rofe a taper fpire, of
wood likewife, reaching feventy feet higher, to the vane
of the weather-cock. Near the bell was fixed an iron
hammer to ftrike the hours; and from the tail of the ham-

mer

mer a wire went down through a fmall gimlet-hole in the floor that the bell ftood upon, and through a fecond floor in like manner ; then horizontally under and near the plaiftered cieling of that fecond floor, till it came near a plaiftered wall ; then down by the fide of that wall to a clock, which ftood about twenty feet below the bell. The wire was not bigger than a common knitting needle. The fpire was fplit all to pieces by the lightning, and the parts flung in all directions over the fquare in which the church ftood, fo that nothing remained above the bell.

The lightning paffed between the hammer and the clock in the above-mentioned wire, without hurting either of the floors, or having any effect upon them, (except making the gimlet-holes, through which the wire paffed, a little bigger,) and without hurting the plaiftered wall, or any part of the building, fo far as the aforefaid wire and the pendulum wire of the clock extended ; which latter wire was about the thicknefs of a goofe-quill. From the end of the pendulum, down quite to the ground, the building was exceedingly rent and damaged, and fome ftones in the foundation-wall torn out, and thrown to the diftance of twenty or thirty feet. No part of the aforementioned long fmall wire, between the clock and the hammer, could be found, except about two inches that hung to the tail of the hammer, and about as much that was faftened to the clock ; the reft being exploded, and its particles diffipated in fmoke and air, as gunpowder is by common fire, and had only left a black fmutty track on

Z

the

the plaiftering, three or four inches broad, darkeft in the middle, and fainter toward the edges, all along the cieling, under which it paffed, and down the wall. Thefe were the effects and appearances; on which I would only make the few following remarks, *viz.*

1. That lightning, in its paffage through a building, will leave wood to pafs as far as it can in metal, and not enter the wood again till the conductor of metal ceafes.

And the fame I have obferved in other inftances, as to walls of brick or ftone.

2. The quantity of lightning that paffed through this fteeple muft have been very great, by its effects on the lofty fpire above the bell, and on the fquare tower all below the end of the clock pendulum.

3. Great as this quantity was, it was conducted by a fmall wire and a clock pendulum, without the leaft damage to the building fo far as they extended.

4. The pendulum rod being of a fufficient thicknefs, conducted the lightning without damage to itfelf; but the fmall wire was utterly deftroyed.

5. Though the fmall wire was itfelf deftroyed, yet it had conducted the lightning with fafety to the building.

6. And from the whole it feems probable, that if even fuch a fmall wire had been extended from the fpindle of the vane to the earth, before the ftorm, no damage would have been done to the fteeple by that ftroke of lightning, though the wire itfelf had been deftroyed.

LET-

L E T T E R XIII.

T O

PETER COLLINSON, *Esq*; F. R. S. at *London*.

Dear Friend,   *Philadelphia, Nov.* 23, 1753.

IN my laft, *via Virginia*, I promifed to fend you per next fhip, a fmall philofophical pacquet: But now having got the materials (old letters and rough drafts) before me, I fear you will find it a great one. Neverthelefs, as I am like to have a few days leifure before this fhip fails, which I may not have again in a long time, I fhall tranfcribe the whole, and fend it; for you will be under no neceffity of reading it all at once, but may take it a little at a time, now and then of a winter evening. When you happen to have nothing elfe to do (if that ever happens,) it may afford you fome amufement *.

B. F.

* Thefe Letters and Papers are a Philofophical Correfpondence between Mr *Franklin* and fome of his *American* Friends. Mr *Collinfon* communicated them to the Royal Society, where they were read at different meetings during the year 1756. But Mr *Franklin* having particularly requefted that they might not be printed, none of them were inferted in the Tranfactions. Mr. *F*. had at that time an intention of revifing them, and purfuing fome of the enquiries farther; but finding that he is not like to have fufficient leifure, he has at length been induced, imperfect as they are, to permit their publication, as fome of the hints they contain may poffibly be ufeful to others in their philofophical refearches.

Ex-

Extract of a Letter from a Gentleman in Boston, *to* Benjamin Franklin, *Efq; concerning the* crooked Direction, *and the* Source *of* Lightning.

S I R,   *Bofton, Dec.* 21, 1751.

THE experiments Mr *K.* has exhibited here, have been greatly pleafing to all forts of people that have feen them; and I hope, by the time he returns to *Philadelphia*, his tour this way will turn to good account. His experiments are very curious, and I think prove moft effectually your doctrine of Electricity; that it is a real element, annexed to, and diffufed among all bodies we are acquainted with; that it differs in nothing from lightning, the effects of both being fimilar, and their properties, fo far as they are known, the fame, &c.

The remarkable effect of lightning on iron, lately difcovered, in giving it the magnetic virtue, and the fame effect produced on fmall needles by the electrical fire, is a further and convincing proof that they are both the fame element; but, which is very unaccountable, Mr *K.* tells me, it is neceffary to produce this effect, that the direction of the needle and the electric fire fhould be North and South; from either to the other, and that juft fo far as they deviate therefrom, the magnetic power in the needle is lefs, till their direction being at right angles with the North and South, the effect entirely ceafes. We made at *Fancuil* Hall,

Hall, where Mr *K——'s* apparatus is, feveral experiments to give fome fmall needles the magnetic virtue ; previoufly examining, by putting them in water, on which they will be fupported, whether or not they had any of that virtue ; and I think we found all of them to have fome fmall degree of it, their points turning to the North: We had nothing to do then but to invert the poles, which accordingly was done, by fending through them the charge of two large glafs jars; the eye of the needle turning to the North, as the point before had done ; that end of the needle which the fire is thrown upon, Mr *K*. tells me always points to the North.

The electrical fire paffing through air has the fame crooked direction as lightning. * This appearance I endeavour to account for thus. Air is an electric *per fe*, therefore there muft be a mutual repulfion betwixt air and the electrical fire. A column or cylinder of air having the diameter of its bafe equal to the diameter of the electrical fpark, intervenes that part of the body which the fpark is taken from, and of the body it aims at. The fpark acts upon this column, and is acted upon by it, more ftrongly than any other neighbouring portion of air.

The column being thus acted upon, becomes more denfe, and being more denfe, repels the fpark more ftrongly; its repellency being in proportion to its denfity: Having acquired, by being condenfed, a degree of repellency greater than its natural, it turns the fpark out of its ftrait courfe; the neighbouring air which muft be lefs denfe, and therefore

* This is moft eafily obferved in large ftrong fparks taken at fome inches diftance.

has

has a fmaller degree of repellency, giving it a more ready paffage.

The fpark having taken a new direction, muft now act on, or moft ftrongly repel the column of air which lies in that direction, and confequently muft condenfe that column in the fame manner as the former, when the fpark muft again change its courfe, which courfe will be thus repeatedly changed, till the fpark reaches the body that attracted it.

To this account one objection occurs; that as air is very fluid and elaftic, and fo endeavours to diffufe itfelf equally, the fuppofed acccumulated air within the column aforefaid, would be immediately diffufed among the contiguous air, and circulate to fill the fpace it was driven from; and confequently that the faid column, on the greater denfity of which the phenomenon is fuppofed to depend, would not repel the fpark more ftrongly than the neighbouring air.

This might be an objection, if the electrical fire was as fluggifh and inactive as air. Air takes a fenfible time to diffufe itfelf equally, as is manifeft from winds which often blow for a confiderable time together from the fame point, and with a velocity even in the greateft ftorms, not exceeding, as it is faid, fixty miles an hour: But the electric fire feems propagated inftantaneoufly, taking up no perceptible time in going very great diftances. It muft then be an inconceivably fhort time in its progrefs from an electrified to an unelectrified body, which, in the prefent cafe, can be but a few inches apart: But this fmall portion of

time

time is not fufficient for the elafticity of the air to exert itfelf, and therefore the column aforefaid muft be in a denfer ftate than its neighbouring air.

About the velocity of the electric fire more is faid below, which perhaps may more fully obviate this objection. But let us have recourfe to experiments. Experiments will obviate all objections, or confound the hypothefis. The electric fpark, if the foregoing be true, will pafs through a vacuum in a right line. To try this, let a wire be fixed perpendicularly on the plate of an air pump, having a leaden ball on its upper end; let another wire paffing through the top of a receiver, have on each end a leaden ball; let the leaden balls within the receiver, when put on the air pump, be within two or three inches of each other : The receiver being exhaufted, the fpark given from a charged vial to the upper wire, will pafs through rarified air, nearly approaching to a vacuum, to the lower wire, and I fuppofe in a right line, or nearly fo; the fmall portion of air remaining in the receiver, which cannot be entirely exhaufted, may poffibly caufe it to deviate a little, but perhaps not fenfibly, from a right line. The fpark alfo might be made to pafs through air greatly condenfed, which perhaps would give a ftill more crooked direction. I have not had opportunity to make any experiments of this fort, not knowing of an air-pump nearer than *Cambridge*, but you can eafily make them. If thefe experiments anfwer, I think the crooked direction of lightning will be alfo accounted for.

With

With refpect to your Letters on Electricity, * * *
* * * * * * * * * * Your Hypothefis in particular for explaining the phænomena of
lightning is very ingenious. That fome clouds are highly
charged with electrical fire, and that their communicating
it to thofe that have lefs, to mountains and other emi-
nencies, makes it vifible and audible, when it is denomina-
ted lightning and thunder, is highly probable : But that
the fea, which you fuppofe the grand fource of it, can collect it, I think admits of a doubt : For though the fea be
compofed of falt and water, an electric *per fe* and non-
electric, and though the friction of electrics *per fe* and non-
electrics, will collect that fire, yet it is only under certain
circumftances, which water will not admit. For it feems
neceffary, that the electrics *per fe* and non-electrics rubbing
one another, fhould be of fuch fubftances as will not adhere
to, or incorporate with each other. Thus a glafs or fulphur fphere turned in water, and fo a friction between
them, will not collect any fire; nor, I fuppofe, would a
fphere of falt revolving in water; the water adhering to,
or incorporating with thofe electrics *per fe*. But granting
that the friction between falt and water would collect the
electrical fire, that fire, being fo extreamly fubtil and active,
would be immediately communicated, either to thofe lower
parts of the fea from which it was drawn, and fo only perform quick revolutions; or be communicated to the adjacent iflands or continent, and fo be diffufed inftantaneoufly through the general mafs of the earth. I fay inftantaneoufly,

neoufly, for the greateft diftances we can conceive within the limits of our globe, even that of the two moft oppofite points, it will take no fenfible time in paffing through : And therefore it feems a little difficult to conceive how there can be any accumulation of the electrical fire upon the furface of the fea, or how the vapours arifing from the fea, fhould have a greater fhare of that fire than other vapours.

That the progrefs of the electrical fire is fo amazingly fwift, feems evident from an experiment you yourfelf (not out of choice) made, when two or three large glafs jars were difcharged through your body. You neither heard the crack, was fenfible of the ftroke, nor, which is more extraordinary, faw the light; which gave you juft reafon to conclude, that it was fwifter than found, than animal fenfation, and even light itfelf. Now light, (as aftronomers have demonftrated) is about fix minutes paffing from the fun to the earth; a diftance, they fay, of more than eighty millions of miles. The greateft rectilinear diftance within the compafs of the earth, is about eight thoufand miles, equal to its diameter. Suppofing then, that the velocity of the electric fire be the fame as that of light, it will go through a fpace equal to the earth's diameter in about $\frac{2}{60}$ of one fecond of a minute. It feems inconceivable then, that it fhould be accumulated upon the fea, in its prefent ftate, which, as it is a non-electric, muft give the fire an inftantaneous paffage to the neighbouring fhores, and they convey it to the general mafs of

A a   the

the earth. But fuch accumulation feems ftill more inconceivable when the electrical fire has but a few feet depth of
water to penetrate, to return to the place, from whence it
is fuppofed to be collected.

Your thoughts upon thefe remarks I fhall receive with
a great deal of pleafure. I take notice that in the printed
copies of your letters feveral things are wanting which are
in the manufcript you fent me. I underftand by your fon,
that you had writ, or was writing, a paper on the effect
of the electrical fire on loadftones, needles, &c. which I
would afk the favour of a copy of, as well as of any other
papers on Electricity, written fince I had the manufcript,
for which I repeat my obligations to you.

I am, &c.

J. B.

LETTER XIV.

FROM

BENJ. FRANKLIN, *Esq*; of *Philadelphia.*

Philadelphia, Jan. 24, 1752.

S I R,

Read at the Roy-
al Society *May*
27, 1756. I AM glad to learn, by your favour of the 21ſt paſt, that Mr *Kinnerſley's* lectures have been acceptable to the Gentlemen of *Boſton*, and are like to prove ſerviceable to himſelf.

I thank you for the countenance and encouragement you have ſo kindly afforded my fellow-citizen.

I ſend you encloſed an extract of a letter containing the ſubſtance of what I obſerved concerning the communication of magnetiſm to needles by Electricity. The minutes I took at the time of the experiments, are miſlaid. I am very little acquainted with the nature of magnetiſm. Dr *Gawin Knight*, inventor of the ſteel magnets, has wrote largely on that ſubject, but I have not yet had leiſure to peruſe his writings with the attention neceſſary to become maſter of his doctrine.

A a 2 Your

Your explication of the crooked direction of lightning, appears to me both ingenious and folid. When we can account as fatisfactorily for the electrification of clouds, I think that branch of Natural Philofophy will be nearly compleat.

The air, undoubtedly, obftructs the motion of the electric fluid. Dry air prevents the diffipation of an electric atmofphere, the denfer the more, as in cold weather. I queftion whether fuch an atmofphere can be retained by a body *in vacuo*. A common electrical vial requires a non-electric communication from the wire to every part of the charged glafs ; otherwife, being dry and clean, and filled with air only, it charges flowly, and difcharges gradually, by fparks, without a fhock : But, exhaufted of air, the communication is fo open and free between the inferted wire and furface of the glafs, that it charges as readily, and fhocks as fmartly as if filled with water : And I doubt not, but that in the experiment you propofe, the fparks would not only be near ftrait *in vacuo*, but ftrike at a greater diftance than in the open air, though perhaps there would not be a loud explofion. As foon as I have a little leifure, I will make the experiment, and fend you the refult.

My fuppofition that the fea might poffibly be the grand fource of lightning, arofe from the common obfervation of its luminous appearance in the night, on the leaft motion ; an appearance never obferved in frefh water. Then I knew that the electric fluid may be pumped up out of the earth, by the friction of a glafs globe, on a non-electric cu-

fhion ;

fhion ; and that, notwithftanding the furprizing activity and fwiftnefs of that fluid, and the non electric communication between all parts of the cufhion and the earth, yet quantities would be fnatch'd up by the revolving furface of the globe, thrown on the prime conductor, and diffipated in air. How this was done, and why that fubtile active fpirit did not immediately return again from the globe, into fome part or other of the cufhion, and fo into the earth, was difficult to conceive ; but whether from its being oppofed by a current fetting upwards to the cufhion, or from whatever other caufe, that it did not fo return was an evident fact. Then I confidered the feparate particles of water as fo many hard fpherules, capable of touching the falt only in points, and imagined a particle of falt could therefore no more be wet by a particle of water, than a globe by a cufhion ; that there might therefore be fuch a friction between thefe originally conftituent particles of falt and water, as in a fea of globes and cufhions ; that each particle of water on the furface might obtain from the common mafs, fome particles of the univerfally diffufed, much finer, and more fubtil electric fluid, and forming to itfelf an atmofphere of thofe particles, be repelled from the then generally electrified furface of the fea, and fly away with them into the air. I thought too, that poffibly the great mixture of particles electric *per fe*, in the ocean water, might, in fome degree, impede the fwift motion and diffipation of the electric fluid through it to the fhores, &c.—But having fince found, that falt in the water of an electric vial, does not leffen the fhock ;

and

and having endeavoured in vain to produce that luminous appearance from a mixture of falt and water agitated ; and obferved, that even the fea-water will not produce it after fome hours ftanding in a bottle ; I fufpect it to proceed from fome principle yet unknown to us (which I would gladly make fome experiments to difcover, if I lived near the fea) and I grow more doubtful of my former fuppofiti-on, and more ready to allow weight to that objection (drawn from the activity of the electric fluid, and the rea-dinefs of water to conduct) which you have indeed ftated with great ftrength and clearnefs.

In the mean time, before we part with this hypothefis, let us think what to fubftitute in its place. I have fome-times queried whether the friction of the air, an electric *per fe*, in violent winds, among trees, and againft the fur-face of the earth, might not pump up, as fo many glafs globes, quantities of the electric fluid, which the rifing va-pours might receive from the air, and retain in the clouds they form ? on which I fhould be glad to have your fenti-ments. An ingenious friend of mine fuppofes the land-clouds more likely to be electrified than the fea-clouds. I fend his letter for your perufal, which pleafe to return me.

I have wrote nothing lately on Electricity, nor obferved any thing new that is material, my time being much taken up with other affairs. Yefterday I difcharged four jars through a fine wire, tied up between two ftrips of glafs : The wire was in part melted, and the reft broke into fmall pieces, from half an inch long, to half a quarter of an inch.

My

My globe raifes the electric fire with greater eafe, in much greater quantities, by the means of a wire extended from the cufhion, to the iron pin of a pump handle behind my houfe, which communicates by the pump fpear with the water in the well.

By this poft I fend to * * * *, who is curious in that way, fome meteorological obfervations and conjectures, and defire him to communicate them to you, as they may afford you fome amufement, and I know you will look over them with a candid eye. By throwing our occafional thoughts on paper, we more readily difcover the defects of our opinions, or we digeft them better, and find new arguments to fupport them. This I fometimes practice ; but fuch pieces are fit only to be feen by friends.

I am, &c.

B. F.

L E T-

LETTER XV.

From J. B. Efq; of *Boſton*,

TO

BENJAMIN FRANKLIN, Efq; at *Philadelphia*.

S I R,   *Boſton, March* 2, 1752.

Read at the Royal Society *June* 3, 1756.

I Have received your favour of the 24th of *January* paſt, incloſing an extract from your letter to Mr *Collinſon*, and * * *'s letter to yourſelf, which I have read with a great deal of pleaſure, and am much obliged to you for. Your extract confirms a correction Mr *Kinnerſley* made a few days ago, of a miſtake I was under reſpecting the polarity given to needles by the electrical fire, " that the end which " receives the fire always points North;" and, " that the " needle being ſituated Eaſt and Weſt, will not have a po- " lar direction." You find, however, the polarity ſtrong- eſt when the needle is ſhocked lying North and South ; weakeſt when lying Eaſt and Weſt ; which makes it pro- bable that the communicated magnetiſm is leſs, as the nee-

dle

dle varies from a North and South fituation. As to the
needle of Capt. *Waddel*'s compafs, if its polarity was reverfed by the lightning, the effect of lightning and Electricity,
in regard of that, feems diffimilar ; for a magnetic needle
in a North and South fituation (as the compafs needle was)
inftead of having its power reverfed, or even diminifhed,
would have it confirmed or increafed by the electric fire.
But perhaps the lightning communicated to fome nails in
the binnacle (where the compafs is placed) the magnetic
virtue, which might difturb the compafs.

This I have heard was the cafe ; if fo, the feeming diffi-
milarity vanifhes : But this remarkable circumftance (if it
took place) I fhould think would not be omitted in Capt.
Waddel's account.

I am very much pleafed that the explication I fent you,
of the crooked direction of lightning, meets with your ap-
probation.

As to your fuppofition about the fource of lightning, the
luminous appearance of the fea in the night, and the fimi-
litude between the friction of the particles of falt and water, as you confidered them in their original feparate ftate,
and the friction of the globe and cufhion, very naturally led
you to the ocean, as the grand fource of lightning : But the
activity of lightning, or the electric element, and the fitnefs
of water to conduct it, together with the experiments you
mention of falt and water, feem to make againft it, and to
prepare the way for fome other hypothefis. Accordingly
you propofe a new one, which is very curious, and not fo

B b   liable,

liable, I think, to objections as the former. But there is not as yet, I believe, a fufficient variety of experiments to eftablifh any theory though this feems the moft hopeful of any I have heard of.

The effect which the difcharge of your four glafs jars had upon a fine wire, tied between two ftrips of glafs, puts me in mind of a very fimilar one of lightning, that I ob-ferved at *New-York, October* 1750, a few days after I left *Philadelphia.* In company with a number of Gentlemen, I went to take a view of the city from the *Dutch* church fteeple, in which is a clock about twenty or twenty-five feet below the bell. From the clock went a wire through two floors, to the clock-hammer near the bell, the holes in the floor for the wire being perhaps about a quarter of an inch diameter. We were told, that in the fpring of 1750, the lightning ftruck the clock-hammer, and defcended a-long the wire to the clock, melting in its way feveral fpots of the wire, from three to nine inches long, through one-third of its fubftance, till coming within a few feet of the lower end, it melted the wire quite through, in feveral pla-ces, fo that it fell down in feveral pieces; which fpots and pieces we faw. When it got to the end of the wire, it flew off to the hinge of a door, fhattered the door, and dif-fipated. In its paffage through the holes of the floors it did not do the leaft damage, which evidences that wire is a good conductor of lightning (as it is of Electricity) pro-vided it be fubftantial enough, and might, in this cafe, had

it

it been continued to the earth, have conducted it without damaging the building *.

Your information about your globe's raising the electric fire in greater quantities, by means of a wire extended from the cushion to the earth, will enable me, I hope, to remedy a great inconvenience I have been under, to collect the fire with the electrifying glafs I use, which is fixed in a very dry room, three ftories from the ground. When you fend your meteorological obfervations to * * * *, I hope I fhall have the pleafure of feeing them

I am, &c.

J. B.

* The wire mentioned in this account was re-placed by a fmall brafs chain. In the fummer of 1763, the lightning again ftruck that fteeple, and from the clock-hammer near the bell, it purfued the chain as it had before done the wire, went off to the fame hinge, and again fhattered the fame door. In its paffage through the fame holes of the fame floors, it did no damage to the floors, nor to the building during the whole extent of the chain. But the chain itfelf was deftroyed, being partly fcattered about in fragments of two or three links melted and ftuck together, and partly blown up or reduced to fmoke, and diffipated.—[See an account of the fame effect of lightning on a wire at *Newbury*, p. 163.] The fteeple, when repair'd, was guarded by an iron conductor, or rod, extending from the foot of the vane-fpindle down the outfide of the building, into the earth.—The newspapers have mentioned, that in 1765, the lightning fell a third time on the fame fteeple, and was fafely conducted by the rod; but the particulars are not come to hand.

*Phyſical and Meteorological Obſervations, Con-
jectures, and Suppoſitions ; by* B. F.

Read at the Royal
Society, *June*
3, 1756. THE particles of air are kept at a diſtance from each other by their mu-
tual repulſion.

Every three particles, mutually and equally repelling each
other, muſt form an equilateral triangle.

All the particles of air gravitate towards the earth, which
gravitation compreſſes them, and ſhortens the ſides of the
triangles, otherwiſe their mutual repellency would force
them to greater diſtances from each other.

Whatever particles of other matter, (not endued with
that repellency) are ſupported in air, muſt adhere to the
particles of air, and be ſupported by them ; for in the vacancies there is nothing they can reſt on.

Air and water mutually attract each other. Hence water
will diſſolve in air, as ſalt in water.

The ſpecific gravity of matter is not altered by dividing
the matter, though the ſuperficies be increaſed. Sixteen
leaden bullets, of an ounce each, weigh as much in water
as one of a pound, whoſe ſuperficies is leſs.

Therefore the ſupporting of ſalt in water is not owing
to its ſuperficies being increaſed.

A lump of ſalt, tho' laid at reſt at the bottom of a veſſel of water, will diſſolve therein, and its parts move every
way, till equally diffuſed in the water ; therefore there is

a mu-

a mutual attraction between water and falt. Every particle of water affumes as many of falt as can adhere to it ;
when more is added, it precipitates, and will not remain fufpended.

Water, in the fame manner, will diffolve in air, every
particle of air affuming one or more particles of water.
When too much is added, it precipitates in rain.

But there not being the fame contiguity between the
particles of air as of water, the folution of water in air is
not carried on without a motion of the air, fo as to caufe a
frefh acceffion of dry particles.

Part of a fluid, having more of what it diffolves, will communicate to other parts that have lefs. Thus very falt water
coming in contact with frefh, communicates its faltnefs till
all is equal, and the fooner if there is a little motion of
the water.

Even earth will diffolve, or mix with air. A ftroke
of a horfe's hoof on the ground, in a hot dufty road,
will raife a cloud of duft, that fhall, if there be a light
breeze, expand every way, till, perhaps, near as big as a
common houfe. It is not by mechanical motion communicated to the particles of duft by the hoof, that they fly fo
far, nor by the wind that they fpread fo wide : But the air
near the ground, more heated by the hot duft ftruck into it,
is rarified and rifes, and in rifing mixes with the cooler air,
and communicates of its duft to it, and it is at length fo
diffufed as to become invifible. Quantities of duft are thus
carried up in dry feafons : Showers wafh it from the air,

and

and bring it down again. For water attracting it ftronger, it quits the air, and adheres to the water.

Air fuffering continual changes in the degrees of its heat, from various caufes and circumftances, and, confequently, changes in its fpecific gravity, muft therefore be in continual motion.

A fmall quantity of fire mixed with water (or degree of heat therein) fo weakens the cohefion of its particles, that thofe on the furface eafily quit it, and adhere to the particles of air.

A greater degree of heat is required to break the cohefion between water and air.

Air moderately heated, will fupport a greater quantity of water invifibly than cold air; for its particles being by heat repelled to a greater diftance from each other, thereby more eafily keep the particles of water that are annexed to them from running into cohefions that would obftruct, refract, or reflect the light.

Hence when we breathe in warm air, though the fame quantity of moifture may be taken up from the lungs, as when we breathe in cold air, yet that moifture is not fo vifible.

Water being extremely heated, *i. e.* to the degree of boiling, its particles in quitting it fo repel each other, as to take up vaftly more fpace than before, and by that repellency fupport themfelves. expelling the air from the fpace they occupy. That degree of heat being leffened, they again mutually attract, and having no air-particles mixed to

ad-

adhere to, by which they might be supported and kept at a distance, they instantly fall, coalesce, and become water again.

The water commonly diffus'd in our atmosphere, never receives such a degree of heat from the sun, or other cause, as water has when boiling ; it is not, therefore, supported by such heat, but by adhering to air.

Water being dissolv'd in, and adhering to air, that air will not readily take up oil, because of the mutual repellency between water and oil.

Hence cold oils evaporate but slowly, the air having generally a quantity of dissolved water.

Oil being heated extremely, the air that approaches its surface will be also heated extremely ; the water then. quitting it, it will attract and carry off oil, which can now adhere to it. Hence the quick evaporation of oil heated to a great degree.

Oil being dissolved in air, the particles to which it adheres will not take up water.

Hence the suffocating nature of air impregnated with burnt grease, as from snuffs of candles, and the like. A certain quantity of moisture should be every moment discharged and taken away from the lungs ; air that has been frequently breath'd, is already overloaded, and, for that reason, can take no more, so will not answer the end. Greasy air refuses to touch it. In both cases suffocation for want of the discharge.

Air will attract and support many other substances.

A par-

A particle of air loaded with adhering water, or any any other matter, is heavier than before, and would defcend.

The atmofphere fuppofed at reft, a loaded defcending particle muft act with a force on the particles   paffes between, or meets with, fufficient to overcome, in fome degree, their mutual repellency, and pufh them nearer to each other.

Thus, fuppofing the particles A B C D, and the others near them, to be at the diftance caufed by their mutual . repellency (confin'd by their common gravity) if A would defcend to E, it muft pafs between B and C; when it comes between B and C it will be nearer to them than before, and muft either have pufh'd them nearer to F and G, contrary to their mutual repellency, or pafs through by a force exceeding its repellency with them.   It then approaches D, and, to move it out of the way, muft act on it with a force fufficient to overcome its repellency with the two next lower particles, by which it is kept in its prefent fituation.

Every particle of air, therefore, will bear any load inferior to the force of thefe repulfions.

Hence the fupport of fogs, mifts, clouds.

Very warm air, clear, though fupporting a very great quantity of moifture, will grow turbid and cloudy on the mixture of a colder air, as foggy turbid air will grow clear by warming.

Thus

Thus the sun shining on a morning fog, dissipates it; clouds are seen to waste in a sun-shiny day.

But cold condenses and renders visible the vapour; a tankard or decanter filled with cold water, will condense the moisture of warm clear air on its outside, where it becomes visible as dew, coalesces into drops, descends in little streams.

The sun heats the air of our atmosphere most near the surface of the earth; for there, besides the direct rays, there are many reflections. Moreover, the earth itself being heated, communicates of its heat to the neighbouring air.

The higher regions having only the direct rays of the sun passing through them, are comparatively very cold. Hence the cold air on the tops of mountains, and snow on some of them all the year, even in the Torrid zone. Hence hail in summer.

If the atmosphere were, all of it (both above and below) always of the same temper as to cold or heat, then the upper air would always be *rarer* than the lower, because the pressure on it is less; consequently lighter, and therefore would keep its place.

But the upper air may be more condensed by cold than the lower air by pressure; the lower more expanded by heat, than the upper for want of pressure. In such case the upper air will become the heavier, the lower the lighter.

The lower region of air being heated and expanded, heaves up, and supports for some time the colder heavier air above, and will continue to support it while the equili-

librium

brium is kept. Thus water is fupported in an inverted o-
pen glafs, while the equilibrium is maintained by the equal
preffure upwards of the air below ; but the equilibrium by
any means breaking, the water defcends on the heavier fide,
and the air rifes into its place.

The lifted heavy cold air over a heated country, becoming by any means unequally fupported, or unequal in its
weight, the heaviest part defcends firft, and the reft follows
impetuoufly. Hence gufts after heats, and hurricanes in
hot climates. Hence the air of gufts and hurricanes cold,
though in hot climes and feafons ; it coming from above.

The cold air defcending from above, as it penetrates our
warm region full of watry particles, condenfes them, renders them vifible, forms a cloud thick and dark, overcaft-
ing fometimes, at once, large and extenfive ; fometimes,
when feen at a diftance, fmall at firft, gradually increafing ;
the cold edge, or furface, of the cloud, condenfing the vapours next it, which form fmaller clouds that join it, increafe its bulk, it defcends with the wind and its acquired
weight, draws nearer the earth, grows denfer with continual additions of water, and difcharges heavy fhowers.

Small black clouds thus appearing in a clear fky, in hot
climates, portend ftorms, and warn feamen to hand their
fails.

The earth turning on its axis in about twenty-fours, the
equatorial parts muft move about fifteen miles in each minute ; in Northern and Southern latitudes this motion is
gradually lefs to the Poles, and there nothing.

If

If there was a general calm over the face of the globe, it muſt be by the air's moving in every part as faſt as the earth or ſea it covers.

He that ſails, or rides, has inſenſibly the ſame degree of motion as the ſhip or coach with which he is connected. If the ſhip ſtrikes the ſhore, or the coach ſtops ſuddenly, the motion continuing in the man, he is thrown forward. If a man were to jump from the land into a ſwift ſailing ſhip, he would be thrown backward (or towards the ſtern) not having at firſt the motion of the ſhip.

He that travels, by ſea or land, towards the equinoctial, gradually acquires motion ; from it, loſes.

But if a man were taken up from latitude 40 (where ſuppoſe the earth's ſurface to move twelve miles *per* minute) and immediately ſet down at the equinoctial, without changing the motion he had, his heels would be ſtruck up, he would fall weſtward. If taken up from the equinoctial, and ſet down in latitude 40, he would fall eaſtward.

The air under the equator, and between the tropics, being conſtantly heated and rarified by the ſun, riſes. Its place is ſupplied by air from Northern and Southern latitudes, which coming from parts where the earth and air had leſs motion, and not ſuddenly acquiring the quicker motion of the equatorial earth, appears an Eaſt wind blowing Weſtward ; the earth moving from Weſt to Eaſt, and ſlipping under the air.

Thus, when we ride in a calm, it ſeems a wind againſt us : If we ride with the wind, and faſter, even that will ſeem a ſmall wind againſt us.

C c 2 The

The air rarified between the Tropics, and rifing, muft, flow in the higher region North and South. Before it rofe, it had acquired the greateft motion the earth's rotation could give it. It retains fome degree of this motion, and defcending in higher latitudes, where the earth's motion is lefs, will appear a Wefterly wind, yet tending towards the equatorial parts, to fupply the vacancy occafioned by the air of the lower regions flowing thitherwards.

Hence our general cold winds are about North-Weft, our fummer cold gufts the fame.

The air in fultry weather, though not cloudy, has a kind of hazinefs in it, which makes objects at a diftance appear dull and indiftinct. This hazinefs is occafioned by the great quantity of moifture equally diffufed in that air. When, by the cold wind blowing down among it, it is con-denfed into clouds, and falls in rain, the air becomes purer and clearer. Hence, after gufts, diftant objects appear dif-tinct, their figures fharply terminated.

Extream cold winds congeal the furface of the earth; by carrying off its fire. Warm winds afterwards blowing over that frozen furface, will be chilled by it. Could that frozen furface be turned under, and a warmer turned up from beneath it, thofe warm winds would not be chilled fo much.

The furface of the earth is alfo fometimes much heated by the fun; and fuch heated furface not being changed, heats the air that moves over it.

Seas, lakes, and great bodies of water, agitated by the winds, continually change furfaces; the cold furface in

win-

winter is turned under, by the rolling of the waves, and a warmer turned up; in fummer, the warm is turned under, and colder turned up. Hence the more equal temper of fea-water, and the air over it. Hence, in winter, winds from the fea feem warm, winds from the land cold. In fummer the contrary.

Therefore the lakes North-Weft of us*, as they are not fo much frozen, nor fo apt to freeze as the earth, rather moderate than increafe the coldnefs of our winter winds.

The air over the fea being warmer, and therefore lighter in winter than the air over the frozen land, may be another caufe of our general N. W. winds, which blow off to fea at right angles from our *North-American* coaft. The warm light fea air rifing, the heavy cold land air preffing into its place.

Heavy fluids defcending, frequently form eddies, or whirlpools, as is feen in a funnel, where the water acquires a circular motion, receding every way from a center, and leaving a vacancy in the middle, greateft above, and leffening downwards, like a fpeaking trumpet, its big end upwards.

Air defcending, or afcending, may form the fame kind of eddies, or whirlings, the parts of air acquiring a circular motion, and receding from the middle of the circle by a centrifugal force, and leaving there a vacancy; if defcending, greateft above, and leffening downwards; if afcending, greateft below, and leffening upwards; like a fpeaking trumpet ftanding its big end on the ground.

When

* In *Penfylvania.*

When the air defcends with violence in fome places, it may rife with equal violence in others, and form both kinds of whirlwinds.

The air in its whirling motion receding every way from the center or axis of the trumpet, leaves there a vacuum ; which cannot be filled through the fides, the whirling air, as an arch, preventing ; it muft then prefs in at the open ends.

The greateft preffure inwards muft be at the lower end, the greateft weight of the furrounding atmofphere being there. The air entering, rifes within, and carries up duft, leaves, and even heavier bodies that happen in its way, as the eddy, or whirl, paffes over land.

If it paffes over water, the weight of the furrounding at-mofphere forces up the water into the vacuity, part of which, by degrees, joins with the whirling air, and adding weight, and receiving accelerated motion, recedes ftill far-ther from the center or axis of the trump, as the preffure lef-fens ; and at laft, as the trump widens, is broken into fmall particles, and fo united with air as to be fupported by it, and become black clouds at the top of the trump.

Thus thefe eddies may be whirlwinds at land, water-fpouts at fea. A body of water fo raifed may be fuddenly let fall, when the motion, *&c.* has not ftrength to fupport it, or the whirling arch is broken fo as to let in the air ; falling in the fea, it is harmlefs, unlefs fhips happen under it. But if in the progreffive motion of the whirl, it has moved from the fea, over the land, and there breaks, fud-den, violent, and mifchievous torrents are the confequences.

L E T-

L E T T E R XVI.

From Dr *** of *Bofton*,

T O

BENJ. FRANKLIN, *Efq*; of *Philadelphia*.

S I R,　　　　　　　　　*Bofton, Aug.* 3, 1752.

THIS comes to you on account of Dr *Douglafs :*
He defired me to write to you for what you know
of the number that died of the inoculation in *Philadelphia,*
telling me he defigned to write fomething on the fmall-
pox fhortly.　We fhall both be obliged to you for a word
on this affair.

The chief particulars of our vifitation, you have in the
public prints.　But the lefs degree of mortality than ufual
in the common way of infection, feems chiefly owing to
the purging method defigned to prevent the fecondary fever ; a method firft begun and carried on in this town, and
with fuccefs beyond expectation.　We loft one in 11 $\frac{1}{6}$, but
had we been experienced in this way, at the firft coming of
the diftemper, probably the proportion had been but one

in

in 13 or 14. In the year 1730 we loft one in nine, which is more favourable than ever before with us. The diftemper pretty much the fame then as now, but fome circumftances not fo kind this time.

If there be any particulars which you want to know, pleafe to fignify what they are, and I fhall fend them.

The number of our inhabitants decreafes*. On a ftrict inquiry, the overfeers of the poor find but 14,190 Whites, and 1,544 Blacks, including thofe abfent, on account of the fmall-pox, many of whom, it is probable, will never return.

I pafs this opportunity without any particulars of my old theme. One thing, however, I muft mention, which is, that perhaps my laft letters contained fomething that feemed to militate with your doctrine of the *Origin, &c.* But my defign was only to relate the phænomena as they appeared to me. I have received fo much light and pleafure from your writings, as to prejudice me in favour of every thing from your hand, and leave me only liberty to obferve, and a power of diffenting when fome great probability might oblige me: And if at any time that be the cafe, you will certainly hear of it.

I am, Sir, &c.

---

* *Bofton* is an old town, and was formerly the feat of all the trade of the country, that was carried on by fea. New towns, and ports, have, of late, divided the trade with it, and diminifhed its inhabitants, though the inhabitants of the country, in general, have greatly increafed.

L E T-

LETTER XVII.

FROM

BENJ. FRANKLIN, *Esq;* of *Philadelphia.*

To Doctor —— of *Boston.*

S I R,   *Philadelphia, Aug.* 13, 1752.

I Received your favour of the 3d instant. Some time last winter I procured from one of our physicians an account of the number of persons inoculated during the five visitations of the small-pox we have had in 22 years; which account I sent to Mr *W—— V——*, of your town, and have no copy. If I remember right, the number exceeded 800, and the deaths were but 4. I suppose Mr *V——* will shew you the account, if he ever received it. Those four were all that our doctors allow to have died of the small-pox by inoculation, though I think there were two more of the inoculated who died of the distemper; but the eruptions appearing soon after the operation, it is supposed they had taken the infection before, in the common way.

D d   I shall

I fhall be glad to fee what Dr *Douglafs* may write on the fubject. I have a *French* piece printed at *Paris* 1724, entitled, *Obfervations fur la Saignée du Pied, et fur la Purgation au commencement de la Petite Verole, & Raifons de doubte contre l'Inoculation.*—A letter of the doctor's is mentioned in it. If he or you have it not, and defire to fee it, I'll fend it.—Pleafe to favour me with the particulars of your purging method, to prevent the fecondary fever.

I am indebted for your preceding letter, but bufinefs fometimes obliges one to poftpone philofophical amufements. Whatever I have wrote of that kind, are really, as they are entitled, but *Conjectures* and *Suppofitions* ; which ought always to give place, when careful obfervation militates againft them. I own I have too ftrong a penchant to the building of hypcthefes ; they indulge my natural indolence: I wifh I had more of your patience and accuracy in making obfervations, on which, alone, true Philfophy can be founded. And, I affure you, nothing can be be more obliging to me, than your kind communication of thofe you make, however they may difagree with my pre-conceived notions.

I am forry to hear that the number of your inhabitants decreafes. I fome time fince, wrote a fmall paper of *Thoughts on the peopling of Countries,* which, if I can find, I will fend you, to obtain your fentiments. The favourable opinion you exprefs of my writings, may, you fee, occafion you more trouble than you expected from, Sir,

Yours, &c.   B. F.

Ob-

OBSERVATIONS *concerning the Increase of Mankind, peopling of Countries,* &c. *Written in* Penfilvania, 1751.

1. TABLES of the proportion of marriages to births, of deaths to births, of marriages to the numbers of inhabitants, &c. formed on obfervations made upon the bills of mortality, chriftenings, &c. of populous cities, will not fuit countries; nor will tables formed on obfervations made on full fettled old countries, as *Europe,* fuit new countries, as *America.*

2. For people increafe in proportion to the number of marriages, and that is greater in proportion to the eafe and convenience of fupporting a family. When families can be eafily fupported, more perfons marry, and earlier in life.

3. In cities, where all trades, occupations, and offices are full, many delay marrying, till they can fee how to bear the charges of a family; which charges are greater in cities, as luxury is more common; many live fingle during life, and continue fervants to families, journeymen to trades, &c. Hence cities do not, by natural generation, fupply themfelves with inhabitants; the deaths are more than the births.

4. In countries full fettled, the cafe muft be nearly the fame; all lands being occupied and improved to the heighth; thofe who cannot get land, muft labour for others that have it; when labourers are plenty, their wages

D d 2
will

will be low ; by low wages a family is fupported with difficulty; this difficulty deters many from marriage, who,
therefore, long continue fervants, and fingle.—Only as the
cities take fupplies of people from the country, and thereby make a little more room in the country, marriage is a
little more encouraged there, and the births exceed the
deaths.

5. Great part of *Europe* is full fettled with hufbandmen,
manufacturers, &c. and therefore cannot now much encreafe in people. *America* is chiefly occupied by *Indians*,
who fubfift moftly by hunting——But as the hunter, of all
men, requires the greateft quantity of land from whence to
draw his fubfiftence, (the hufbandman fubfifting on much
lefs, the gardener on ftill lefs, and the manufacturer requiring leaft of all) the *Europeans* found *America* as fully fettled as it well could be by hunters ; yet thefe having large
tracts, were eafily prevailed on to part with portions of
territory to the new comers, who did not much interfere
with the natives in hunting, and furnifhed them with many things they wanted.

6. Land being thus plenty in *America*, and fo cheap as
that a labouring man that underftands hufbandry, can, in
a fhort time, fave money enough to purchafe a piece of
new land, fufficient for a plantation, whereon he may fubfift a family ; fuch are not afraid to marry ; for if they e-
ven look far enough forward to confider how their children, when grown up, are to be provided for, they fee that

more

more land is to be had at rates equally eafy, all circumftances confidered.

7. Hence marriages in *America* are more general, and more generally early than in *Europe*. And if it is reckoned there, that there is but one marriage *per Annum* among 100 perfons, perhaps we may here reckon two; and if in *Europe* they have but four births to a marriage, (many of their marriages being late) we may here reckon eight; of which, if one half grow up, and our marriages are made, reckoning one with another, at twenty years of age, our people muft at leaft be doubled every twenty years.

8. But notwithftanding this increafe, fo vaft is the territory of *North-America*, that it will require many ages to fettle it fully; and till it is fully fettled, labour will never be cheap here, where no man continues long a labourer for others, but gets a plantation of his own; no man continues long a journeyman to a trade, but goes among thofe new fettlers, and fets up for himfelf, &c. Hence labour is no cheaper now, in *Penfilvania*, than it was thirty years ago, though fo many thoufand labouring people have been imported from *Germany* and *Ireland*.

9. The danger, therefore, of thefe colonies interfereing with their mother country in trades that depend on labour, manufactures, &c. is too remote to require the attention of *Great-Britain*.

10. But in proportion to the increafe of the colonies, a vaft demand is growing for *Britifh* manufactures; a glorious market, wholly in the power of *Britain*, in which

foreigners

foreigners cannot interfere, which will increase, in a short time, even beyond her power of supplying, though her whole trade should be to her colonies. * * * *.

12. 'Tis an ill-grounded opinion, that by the labour of slaves, *America* may possibly vie in cheapness of manufactures with *Britain*. The labour of slaves can never be so cheap here, as the labour of working men is in *Britain*. Any one may compute it. Interest of money is in the colonies from 6 to 10 *per Cent*. Slaves, one with another, cost 30*l*. sterling *per* head. Reckon then the interest of the first purchase of a slave, the insurance or risque on his life, his cloathing and diet, expences in his sickness, and loss of time, loss by his neglect of business, (neglect is natural to the man who is not to be benefited by his own care or diligence) expence of a driver to keep him at work, and his pilfering from time to time, almost every slave being, from the nature of slavery, a thief; and compare the whole amount with the wages of a manufacturer of iron or wool in *England*, you will see that labour is much cheaper there, than it ever can be by negroes here. Why then will *Americans* purchase slaves? Because slaves may be kept as long as a man pleases, or has occasion for their labour; while hired men are continually leaving their master (often in the midst of his business) and setting up for themselves. § 8.

13. As the increase of people depends on the encouragement of marriages, the following things must diminish a nation, *viz.* 1. The being conquered; for the con-

querors

querors will engrofs as many offices, and exact as much tribute or profit on the labour of the conquered. as will maintain them in their new eftablifhment ; and this diminifhing the fubfiftence of the natives, difcourages their marriages, and fo gradually diminifhes them, while the foreigners increafe. 2. Lofs of territory. Thus the *Britons* being driven into *Wales*, and crouded together in a barren country, infufficient to fupport fuch great numbers, diminifhed, till the people bore a proportion to the produce, while the *Saxons* increafed on their abandoned lands, 'till the ifland became full of *Englifh.* And, were the *Englifh* now driven into *Wales* by fome foreign nation, there would, in a few years, be no more *Englifhmen* in *Britain*, than there are now people in *Wales.* 3. Lofs of trade. Manufactures exported, draw fubfiftence from foreign countries for numbers ; who are thereby enabled to marry and raife families. If the nation be deprived of any branch of trade, and no new employment is found for the people occupied in that branch, it will foon be deprived of fo many people. 4. Lofs of food. Suppofe a nation has a fifhery, which not only employs great numbers, but makes the food and fubfiftence of the people cheaper : if another nation becomes mafter of the feas, and prevents the fifhery, the people will diminifh in proportion as the lofs of employ, and dearnefs of provifion makes it more difficult to fubfift a family. 5. Bad government and infecure property. People not only leave fuch a country, and fettling abroad incorporate with other nations, lofe

their

their native language, and become foreigners; but the induftry of thofe that remain being difcouraged, the quantity of fubfiftence in the country is leffened, and the fupport of a family becomes more difficult. So heavy taxes
tend to diminifh a people. 6. The introduction of flaves.
The negroes brought into the *Englifh* fugar iflands, have
greatly diminifhed the Whites there; the poor are by this
means deprived of employment, while a few families acquire vaft eftates, which they fpend on foreign luxuries,
and educating their children in the habit of thofe luxuries ; the fame income is needed for the fupport of one,
that might have maintained one hundred. The whites,
who have flaves not labouring, are enfeebled, and therefore not fo generally prolific ; the flaves being worked too
hard, and ill fed, their conftitutions are broken, and the
deaths among them are more than the births ; fo that a
continual fupply is needed from *Africa*. The northern
colonies having few flaves, increafe in whites. Slaves
alfo pejorate the families that ufe them ; the white children become proud, difgufted with labour, and being e-
ducated in idlenefs, are rendered unfit to get a living by
induftry.

14. Hence the prince that acquires new territory, if he
finds it vacant, or removes the natives to give his own
people room ; the legiflator that makes effectual laws for
promoting of trade, increafing employment, improving
land by more or better tillage, providing more food by
fifheries, fecuring property, *&c.* and the man that invents
new

new trades, arts, or manufactures, or new improvements
in husbandry, may be properly called *Fathers of their na-*
tion, as they are the cause of the generation of multi-
tudes, by the encouragement they afford to marriage.

15. As to privileges granted to the married, (such as
the *jus trium liberorum* among the *Romans*) they may
hasten the filling of a country that has been thinned by
war or pestilence, or that has otherwise vacant territory,
but cannot increase a people beyond the means provided
for their subsistence.

16. Foreign luxuries, and needless manufactures, imported and used in a nation, do, by the same reasoning,
increase the people of the nation that furnishes them, and
diminish the people of the nation that uses them.—Laws,
therefore, that prevent such importations, and, on the
contrary, promote the exportation of manufactures to be
consumed in foreign countries, may be called (with respect to the people that make them) *generative laws*, as
by increasing subsistence they encourage marriage. Such
laws likewise strengthen a country doubly, by increasing
its own people, and diminishing its neighbours.

17. Some *European* nations prudently refuse to con-
sume the manufactures of *East-India* :—They should
likewise forbid them to their colonies; for the gain to the
merchant is not to be compared with the loss, by this
means, of people to the nation.

18. Home luxury in the great, increases the nation's
manufacturers employed by it, who are many, and only

E e

tends

tends to diminiſh the families that indulge in it, who are few. The greater the common faſhionable expence of any rank of people, the more cautious they are of marriage. Therefore luxury ſhould never be ſuffered to become common.

19. The great increaſe of offspring in particular families, is not always owing to greater fecundity of nature, but ſometimes to examples of induſtry in the heads, and induſtrious education ; by which the children are enabled to provide better for themſelves, and their marrying early is encouraged from the proſpect of good ſubſiſtence.

20. If there be a ſect, therefore, in our nation, that regard frugality and induſtry as religious duties, and educate their children therein, more than others commonly do ; ſuch ſect muſt conſequently increaſe more by natural generation, than any other ſect in *Britain*.

21. The importation of foreigners into a country that has as many inhabitants as the preſent employments and proviſions for ſubſiſtence will bear, will be in the end no increaſe of people, unleſs the new-comers have more induſtry and frugality than the natives, and then they will provide more ſubſiſtence, and increaſe in the country; but they will gradually eat the natives out.—Nor is it neceſſary to bring in foreigners to fill up any occaſional vacancy in a country ; for ſuch vacancy (if the laws are good, § 14, 16) will ſoon be filled by natural generation. Who can now find the vacancy made in *Sweden, France,* or other warlike nations, by the plague of heroiſm 40 years ago ;

ago; in *France*, by the expulsion of the Protestants; in *England*, by the settlement of her colonies; or in *Guinea*, by a hundred years exportation of slaves, that has blackened half *America* ?—The thinness of the inhabitants in *Spain*, is owing to national pride, and idleness, and other causes, rather than to the expulsion of the *Moors*, or to the making of new settlements.

22. There is, in short, no bound to the prolific nature of plants or animals, but what is made by their crowding and interfering with each other's means of subsistence. Was the face of the earth vacant of other plants, it might be gradually sowed and overspread with one kind only; as for instance, with Fennel; and were it empty of other inhabitants, it might, in a few ages, be replenished from one nation only, as for instance, with *Englishmen*. Thus there are supposed to be now upwards of one million *English* souls in *North-America*, (though it is thought scarce 80,000 have been brought over sea) and yet perhaps there is not one the fewer in *Britain*, but rather many more, on account of the employment the colonies afford to manufacturers at home. This million doubling suppose but once in 25 years, will, in another century, be more than the people of *England*, and the greatest number of *Englishmen* will be on this side the water. What an accession of power to the *British* empire by sea as well as land! What increase of trade and navigation! What numbers of ships and seamen! We have been here but little more than a hundred years, and yet the force of our privateers in the

late

late war, united, was greater, both in men and guns, than that of the whole *Britiſh* navy in Queen *Elizabeth*'s time ——How important an affair, then, to *Britain*, is the preſent treaty * for ſettling the bounds between her colonies, and the *French !* and how careful ſhould ſhe be to ſecure room enough, ſince on the room depends ſo much the increaſe of her people ?

23. In fine, a nation well regulated is like a polypus † ; take away a limb, its place is ſoon ſupplied ; cut it in two, and each deficient part ſhall ſpeedily grow out of the part remaining. Thus if you have room and ſubſiſt-ence enough, as you may, by dividing, make ten poly-puſes out of one, you may, of one, make ten nations, equally populous and powerful ; or, rather, increaſe a na-tion ten fold in numbers and ſtrength. * * * * *

---

* In 1751.   † A water-inſect, well known to Naturaliſts.

LET-

LETTER XVIII.

From Doctor ———, of *Boston,*

T O

BENJAMIN FRANKLIN, Efq; at *Philadelphia.*

S I R,   *Boston, October* 16, 1752.

Read at the Roy-
al Society *June*
3, 1756.

I Find, by a word or two in your laft, that you are willing to be found fault with ; which authorifes me to let you know what I am at a lofs about in your papers, which is only in the article of the water-fpout. I am in doubt, whether water in bulk, or even broken into drops, ever afcends into the region of the clouds, *per vorticem,* (*i. e.*) whether there be, in reality, what I call a direct water-fpout. I make no doubt of direct and inverted whirl-winds; your defcription of them, and the reafon of the thing, are fufficient. I am fenfible, too, that they are very ftrong, and often move confiderable weights. But I have not met with any hiftorical accounts that feem ex-act enough to remove my fcruples concerning the afcent above faid.

Defcend-

Defcending fpouts (as I take them to be) are many times feen, as I take it, in the calms, between the fea and land trade-winds, on the coaft of *Africa*. Thefe contrary winds, or diverging, I can conceive may occafion them, as it were by fuction, making a breach in a large cloud. But I imagine they have, at the fame time, a tendency to hinder any direct or rifing fpout, by carrying off the lower part of the atmofphere, as faft as it begins to rarefy ; and yet fpouts are frequent here, which ftrengthens my opinion, that all of them defcend.

But however this be, I cannot conceive a force producible by the rarification and condenfation of our atmofphere, in the circumftances of our globe, capable of carrying water, in large portions, into the region of the clouds. Suppofing it to be raifed, it would be too heavy to continue the afcent beyond a confiderable height, unlefs parted into fmall drops ; and, even then, by its centrifugal force, from the manner of conveyance, it would be flung out of the circle, and fall fcattered, like rain.

But I need not expatiate on thefe matters to you. I have mentioned my objections, and, as truth is my purfuit, fhall be glad to be informed. I have feen few accounts of thefe whirl, or eddy winds, and as little of the fpouts ; and thefe, efpecially, lame and poor things to obtain any certainty by. If you know any thing determinate that has been obferved, I fhall hope to hear from you ; as alfo of any miftake in my thoughts.

I have

I have nothing to object to any other part of your suppositions; and as to that of the trade-winds, I believe nobody can.

I am, &c.

P. S. The figures in the *Philosophical Transactions* shew, by several circumstances, that they all descended, though the relators seemed to think they took up water.

LETTER XIX.

From Doctor ———, of *Boston*,

T O

Benjamin Franklin, *Esq*; of *Philadelphia*.

S I R,   *Boston, October* 23, 1752.

Read at the Royal Society, *June* 24, 1754. IN the inclosed you have all I have to say of that matter *. It proved longer than I expected, so that I was forced to add a cover to it. I confess it looks like a dispute; but that is quite contrary to my intentions. The sincerity of friendship and esteem were my motives; nor do I doubt

* Water-spouts.

your

your fcrupling the goodnefs of the intention. However, I muft confefs I cannot tell exactly how far I was acted by hopes of better information, in difcovering the whole foundation of my opinion, which, indeed, is but an opinion, as I am very much at a lofs about the validity of the reafons. I have not been able to differ from you in fentiment concerning any thing elfe in your *Suppofitions.* In the prefent cafe I lie open to conviction, and fhall be the gainer when informed. If I am right, you will know that, without my adding any more. Too much faid on a merely fpeculative matter, is but a robbery committed on practical knowledge. Perhaps I am too much pleafed with thefe dry notions: However, by this you will fee that I think it unreafonable to give you more trouble about them, than your leifure and inclination may prompt you to.

I am, &c.

SINCE my laft I confidered, that, as I had begun with the reafons of my diffatisfaction about the afcent of water in fpouts, you would not be unwilling to hear the whole I have to fay, and then you will know what I rely upon.

What occafioned my thinking all fpouts defcend, is, that I found fome did certainly do fo. A difficulty appeared concerning the afcent of fo heavy a body as water, by any force I was apprifed of, as probably fufficient.

And,

And, above all, a view of Mr *Stuart*'s portraits of fpouts, in the *Philofophical Tranfactions.*

Some obfervations on thefe laft, will include the chief part of my difficulties.

Mr *Stuart* has given us the figures of a number obferved by him in the *Mediterranean* : All with fome particulars which make for my opinion, if well drawn.

The great fpattering which relators mention in the water where the fpout defcends, and which appears in all his draughts, I conceive to be occafioned by drops defcending very thick and large into the place.

On the place of this fpattering arifes the appearance of a bufh, into the center of which the fpout comes down. This bufh I take to be formed by a fpray, made by the force of thefe drops, which being uncommonly large, and defcending with unufual force, by a ftream of wind defcending from the cloud with them, increafes the height of the fpray ; which wind being repulfed by the furface of the waters, rebounds and fpreads ; by the firft raifing the fpray higher than otherwife it would go ; and by the laft making the top of the bufh appear to bend outwards (*i. e.*) the cloud of fpray is forc'd off from the trunk of the fpout, and falls backward.

The bufh does the fame, where there is no appearance of a fpout reaching it ; and is depreffed in the middle, where the fpout is expected. This, I imagine, to be from numerous drops of the fpout falling into it, together

with

with the wind I mentioned, by their defcent, which beat back the rifing fpray in the center.

This circumftance, of the bufh bending outwards at the top, feems not to agree with what I call a direct whirlwind, but confiftent with the revers'd ; for a direct one would fweep the bufh inwards ; if, in that cafe, any thing of a bufh would appear.

The pillar of water, as they call it, from its likenefs, I fuppofe to be only the end of the fpout immers'd in the bufh, a little blacken'd by the additional cloud, and, per-haps, appears to the eye beyond its real bignefs, by a refraction in the bufh, and which refraction may be the caufe of the appearance of feparation, betwixt the part in the bufh, and that above it. The part in the bufh is cylindrical, as it is above (*i. e.*) the bignefs the fame from the top of the bufh to the water. Inftead of this fhape, in cafe of a whirlwind, it muft have been pyramidical.

Another thing remarkable, is, the curve in fome of them : This is eafy to conceive, in cafe of defcending parcels of drops through various winds, at leaft till the cloud condenfes fo faft as to come down, as it were, *uno rivo*. But it is harder to me to conceive it in the afcent of water, that it fhould be conveyed along, fecure of not leaking, or often dropping through the under fide, in the prone part : And, fhould the water be conveyed fo fwift-ly, and with fuch force, up into the cloud, as to pre-vent this ; it would, by a natural difpofition to move on

in

in a prefent direction, prefently ftraiten the curve, raifing the fhoulder very fwiftly, till loft in the cloud.

Over every one of *Stuart*'s figures, I fee a cloud: I fuppofe his clouds were firft, and then the fpout ; I do not know whether it be fo with all fpouts, but fuppofe it is. Now, if whirlwinds carried up the water, I fhould expect them in fair weather, but not under a cloud ; as is obfervable of whirlwinds ; they come in fair weather, not under the fhade of a cloud, nor in the night ; fince fhade cools the air : But, on the contrary, violent winds often defcend from the clouds ; ftrong gufts which occupy fmall fpaces ; and from the higher regions, extenfive hurricanes, &c.

Another thing is, the appearance of the fpout *coming from* the cloud. This I cannot account for on the notion of a direct fpout, but, in the real defcending one, it is eafy. I take it, that the cloud begins firft of all to pour out drops at that particular fpot, or *foramen* ; and, when that current of drops increafes, fo as to force down wind and vapour, the-fpout becomes, fo far as that goes, opaque. I take it, that no clouds drop fpouts, but fuch as make very faft, and happen to condenfe in a particular fpot, which, perhaps, is coldeft, and gives a determination downwards, fo as to make a paffage through the fubjacent atmofphere.

If fpouts afcend, it is to carry up the warm rarified air below, to let down all and any that is colder above ; and, if fo, they muft carry it through the cloud they go

into,

into, (for that is cold and denfe, I imagine) perhaps far into the higher region, making a wonderful appearance at a convenient diftance to obferve it, by the fwift rife of a body of vapour, above the region of the clouds. But, as this has never been obferved in any age, if it be fuppofeable that is all.

I cannot learn, by marincrs, that any wind blows towards a fpout more than any other way; but it blows towards a whirlwind, for a large diftance round.

I fuppofe there has been no inftance of the water of a fpout being falt, when coming a-crofs any veffel at fea. I fuppofe, too, that there have been no falt rains; thefe would make the cafe clear.

I fuppofe it is from fome unhappy effects of thefe dangerous creatures of nature, that failors have an univerfal dread on them of breaking in their decks, fhould they come acrofs them.

I imagine fpouts, in cold feafons, as *Gordon's* in the *Downs,* prove the defcent.

Query. Whether there is not always more or lefs cloud, firft, where a fpout appears?

Whether they are not, generally, on the borders of trade-winds; and whether this is for, or againft me?

Whether there be any credible account of a whirlwind's carrying up all the water in a pool, or fmall pond : As when fhoal, and the banks low, a ftrong guft might be fuppofed to blow it all out?

Whether

Whether a violent tornado, of a fmall extent, and o-
ther fudden and ftrong gufts, be not winds from above,
defcending nearly perpendicular; and, whether many that
are called whirlwinds at fea, are any other than thefe;
and fo might be called air-fpouts, if they were objects of
fight?

I overlooked, in its proper place, *Stuart*'s No. 11,
which is curious for its inequalities, and, in particular,
the approach to breaking, which, if it would not be too
tedious, I would have obferv'd a little upon, in my own
way, as, I think, this would argue againft the afcent, *&c.*
but I muft pafs it, not only for the reafon mentioned, but
want of room befides.

As to Mr *Stuart*'s ocular demonftration of the afcent
in his great perpendicular fpout, the only one it appears
in, I fay, as to this, what I have written fuppofes him
miftaken, which, yet, I am far from afferting.

The force of an airy vortex, having lefs influence on
the folid drops of water, than on the interfperfed cloudy
vapours, makes the laft whirl round fwifter, though it de-
foend flower: And this might eafily deceive, without
great care, the moft unprejudiced perfon.

LET-

LETTER XX.

FROM

BENJ. FRANKLIN, *Esq*; of *Philadelphia*.

To Doctor *** of *Boston*.

SIR,   *Philadelphia, Feb.* 4, 1753

Read at the Royal
Society *June*
24, 1756.

I Ought to have written to you, long since, in anfwer to yours of *October* 16, concerning the water-fpout; but bufinefs partly, and partly a defire of procuring further information, by enquiry among my fea-faring acquaintance, induced me to poftpone writing, from time to time, till I am now almoft afhamed to refume the fubject, not knowing but you may have forgot what has been faid upon it.

Nothing, certainly, can be more improving to a fearcher into nature, than objections judicioufly made to his opinion, taken up, perhaps, too haftily: For fuch objections oblige him to re-ftudy the point, confider every circumftance carefully, compare facts, make experiments, weigh arguments, and be flow in drawing conclufions.

And

And hence a sure advantage results; for he either confirms a truth, before too slightly supported; or discovers an error, and receives instruction from the objector.

In this view I consider the objections and remarks you sent me, and thank you for them sincerely: But, how much soever my inclinations lead me to Philosophical inquiries, I am so engaged in business, public and private, that those more pleasing pursuits are frequently interrupted, and the chain of thought, necessary to be closely continued in such disquisitions, so broken and disjointed, that it is with difficulty I satisfy myself in any of them: And I am now not much nearer a conclusion, in this matter of the spout, than when I first read your letter.

Yet, hoping we may, in time, sift out the truth between us, I will send you my present thoughts, with some observations on your reasons, on the accounts in the *Transactions*, and on other relations I have met with. Perhaps, while I am writing, some new light may strike me, for I shall now be obliged to consider the subject with a little more attention.

I agree with you, that, by means of a vacuum in a whirlwind, water cannot be supposed to rise in large masses to the region of the clouds; for the pressure of the surrounding atmosphere could not force it up in a continued body, or column, to a much greater height than thirty feet. But, if there really is a vacuum in the center, or near the axis of whirlwinds, then, I think, water

may

may rife in fuch vacuum to that height, or to lefs height, as the vacuum may be lefs perfect.

I had not read *Stuart*'s account, in the *Tranfactions*, for many years, before the receipt of your letter, and had quite forgot it ; but now, on viewing his draughts, and confidering his defcriptions, I think they feem to favour *my hypothefis*; for he defcribes and draws columns of water, of various heights, terminating abruptly at the top, exactly as water would do, when forced up by the preffure of the atmofphere, into an exhaufted tube.

I muft, however, no longer call it *my hypothefis*, fince I find *Stuart* had the fame thought, though fomewhat obfcurely expreffed, where he fays " he imagines this phæ- " nomenon may be folv'd by fuction (improperly fo call- " ed) or rather pulfion, as in the application of a cup- " ping glafs to the flefh, the air being firft voided by the " kindled flax."

In my paper, I fuppofed a whirlwind and a fpout to be the fame thing, and to proceed from the fame caufe ; the only difference between them being, that the one paffes over land, the other over water. I find, alfo, in the *Tranfactions*, that M. *de la Pryme* was of the fame opinion ; for he there defcribes two fpouts, as he calls them, which were feen at different times, at *Hatfield* in *Yorkſhire*, whofe appearances in the air were the fame with thofe of the fpouts at fea, and effects the fame with thofe of real whirlwinds.

Whirl-

Whirlwinds have, generally, a progreffive, as well as a circular motion; fo had what is called the fpout, at *Topfham*—*(See the account of it in the Tranfactions)*—which alfo appears, by its effects defcribed, to have been a real whirlwind. Water-fpouts have, alfo, a progreffive motion; this is fometimes greater, and fometimes lefs; in fome violent, in others barely perceivable. The whirlwind at *Warrington* continued long in *Acrement-Clofe*.

Whirlwinds generally arife after calms and great heats: The fame is obferved of water-fpouts, which are, therefore, moft frequent in the warm latitudes. The fpout that happened in cold weather, in the *Downs*, defcribed by Mr *Gordon* in the *Tranfactions*, was, for that reafon, thought extraordinary; but he remarks withal, that the weather, though cold when the fpout appeared, was foon after much colder; as we find it, commonly, lefs warm after a whirlwind.

You agree, that the wind blows every way towards a whirlwind, from a large fpace round. An intelligent whaleman of *Nantucket*, informed me, that three of their veffels, which were out in fearch of whales, happening to be becalmed, lay in fight of each other, at about a league diftance, if I remember right, nearly forming a triangle: After fome time, a water-fpout appeared near the middle of the triangle, when a brifk breeze of wind fprung up, and every veffel made fail; and then it appeared to them all, by the fetting of the fails, and the courfe each veffel ftood, that the fpout was to the leeward of every one of

G g   them;

them ; and they all declared it to have been so, when they happened afterwards in company, and came to confer about it. So that in this particular likewise, whirlwinds and water-spouts agree.

But, if that which appears a water-spout at sea, does sometimes, in its progressive motion, meet with and pass over land, and there produce all the phænomena and effects of a whirlwind, it should thence seem still more evident, that a whirlwind and a spout are the same. I send you, herewith, a letter from an ingenious physician of my acquaintance, which gives one instance of this, that fell within his observation.

A fluid, moving from all points horizontally, towards a center, must, at that center, either ascend or descend. Water being in a tub, if a hole be opened in the middle of the bottom, will flow from all sides to the center, and there descend in a whirl. But, air flowing on and near the surface of land or water, from all sides, towards a center, must, at that center, ascend ; the land or water hindering its descent.

If these concentring currents of air be in the upper region, they may, indeed, descend in the spout or whirlwind ; but then, when the united current reached the earth or water, it would spread, and, probably, blow every way from the center. There may be whirlwinds of both kinds, but, from the commonly observed effects, I suspect the rising one to be the most common : When the upper air descends, it is, perhaps, in a greater body, extending

tending wider, as in our thunder-gufts, and without much whirling ; and, when air defcends in a fpout, or whirl-wind, I fhould rather expect it would prefs the roof of a houfe *inwards,* or force *in* the tiles, fhingles, or thatch, force a boat down into the water, or a piece of timber in-to the earth, than that it would lift them up, and carry them away.

It has fo happened, that I have not met with any ac-counts of fpouts, that certainly defcended ; I fufpect they are not frequent. Pleafe to communicate thofe you men-tion. The apparent dropping of a pipe from the clouds towards the earth or fea, I will endeavour to explain here-after.

The augmentation of the cloud, which, as I am in-formed, is generally, if not always the cafe, during a fpout, feems to fhew an afcent, rather than a defcent of the mat-ter of which fuch cloud is compofed ; for a defcending fpout, one would expect, fhould diminifh a cloud. I own, however, that cold air defcending, may, by con-denfing the vapours in a lower region, form and increafe clouds ; which, I think, is generally the cafe in our common thunder-gufts, and, therefore, do not lay great ftrefs on this argument.

Whirlwinds, and fpouts, are not always, though moft commonly, in the day time. The terrible whirlwind which damaged a great part of *Rome, June* 11, 1749, happened in the night of that day. The fame was fup-pofed to have been firft a fpout, for it is faid to be beyond

G g 2 doubt,

doubt, that it gathered in the neighbouring fea, as it could be tracked from *Oftia* to *Rome*. I find this in *Pere Bofchovich's* account of it, as abridg'd in the *Monthly Review* for *December* 1750.

In that account, the whirlwind is faid to have appeared as a very black, long, and lofty cloud, difcoverable, notwithftanding the darknefs of the night, by its continually lightning or emitting flafhes on all fides, pufhing along with a furprizing fwiftnefs, and within three or four feet of the ground. Its general effects on houfes, were, ftripping off the roofs, blowing away chimneys, breaking doors and windows, *forcing up the floors, and unpaving the rooms*, (fome of thefe effects feem to agree well with a fuppofed vacuum in the center of the whirlwind) and the very rafters of the houfes were broke and difperfed, and even hurled againft houfes at a confiderable diftance, &c.

It feems, by an expreffion of *Pere Bofchovich's*, as if the wind blew from all fides towards the whirlwind; for, having carefully obferved its effects, he concludes of all whirlwinds, " that their motion is circular, and their ac-
" tion attractive."

He obferves, on a number of hiftories of whirlwinds, &c. " that a common effect of them is, to carry up into
" the air tiles, ftones, and animals themfelves, which hap-
" pen to be in their courfe, and all kinds of bodies unex-
" ceptionably, throwing them to a confiderable diftance,
" with

" with great impetuofity." Such effects feem to fhew a rifing current of air.

I will endeavour to explain my conceptions of this matter by figures, reprefenting a plan and an elevation of a fpout or whirlwind.

I would only firft beg to be allowed two or three pofitions, mentioned in my former paper.

1. That the lower region of air is often more heated, and fo more rarified, than the upper ; confequently, fpecifically lighter. The coldnefs of the upper region is manifefted by the hail which fometimes falls from it in a hot day.

2. That heated air may be very moift, and yet the moifture fo equally diffus'd and rarified, as not to be vifible, till colder air mixes with it, when it condenfes, and becomes vifible. Thus our breath, invifible in fummer, becomes vifible in winter.

Now, let us fuppofe a tract of land, or fea, of perhaps fixty miles fquare, unfcreened by clouds, and unfanned by winds, during great part of a fummer's day, or, it may be, for feveral days fucceffively, till it is violently heated, together with the lower region of air in contact with it, fo that the faid lower air becomes fpecifically lighter than the fuperincumbent higher region of the atmofphere, in which the clouds commonly float : Let us fuppofe, alfo, that the air furrounding this tract has not been fo much heated during thofe days, and, therefore, remains heavier. The confequence of this fhould be, as I conceive, that the

heated

heated lighter air, being preſſed on all ſides, muſt aſcend, and the heavier deſcend ; and, as this riſing cannot be in all parts, or the whole area of the tract at once, for that would leave too extenſive a vacuum, the riſing will begin preciſely in that column that happens to be the lighteſt, or moſt rarified ; and the warm air will flow horizontally from all points to this column, where the ſeveral currents meeting, and joining to riſe, a whirl is naturally formed, in the ſame manner as a whirl is formed in the tub of water, by the deſcending fluid flowing from all ſides of the tub, to the hole in the center.

And, as the ſeveral currents arrive at this central riſing column, with a conſiderable degree of horizontal motion, they cannot ſuddenly change it to a vertical motion ; therefore, as they gradually, in approaching the whirl, decline from right to curve or circular lines, ſo, having joined the whirl, they *aſcend* by a ſpiral motion ; in the ſame manner as the water *deſcends* ſpirally through the hole in the tub before-mentioned.

Laſtly, as the lower air, and neareſt the ſurface, is moſt rarified by the heat of the ſun, that air is moſt acted on by the preſſure of the ſurrounding cold and heavy air, which is to take its place ; conſequently, its motion towards the whirl is ſwifteſt, and ſo the force of the lower part of the whirl, or trump, ſtrongeſt, and the centrifugal force of its particles greateſt ; and hence the vacuum round the axis of the whirl ſhould be greateſt near the earth or ſea, and be gradually diminiſhed as it ap-

proaches

proaches the region of the clouds, till it ends in a point, as at A in Fig. II. forming a long and sharp cone.

In Fig. I. which is a plan or ground-plat of a whirlwind, the circle V. reprefents the central vacuum.

Between *a a a a* and *b b b b* I suppofe a body of air condenfed ftrongly by the preffure of the currents moving towards it, from all fides without, and by its centrifugal force from within; moving round with prodigious fwiftnefs, (having, as it were, the momenta of all the currents

$$\longrightarrow \longrightarrow \longrightarrow \longrightarrow$$

united in itfelf) and with a power equal to its fwiftnefs and denfity.

It is this whirling body of air between *a a a a* and *b b b b* that rifes fpirally; by its force it tears buildings to pieces, twifts up great trees by the roots, &c. and, by its fpiral motion, raifes the fragments fo high, till the preffure of the furrounding and approaching currents diminifhing, can no longer confine them to the circle; or their own centrifugal force encreafing, grows too ftrong for fuch preffure, when they fly off in tangent lines, as ftones out of a fling, and fall on all fides, and at great diftances.

If it happens at fea, the water under and between *a a a a* and *b b b b* will be violently agitated and driven about, and parts of it raifed with the fpiral current, and thrown about, fo as to form a bufh like appearance.

This circle is of various diameters, fometimes very large.

If the vacuum paffes over water, the water may rife in it in a body, or column, to near the height of thirty-two feet.

If

If it paffes over houfes, it may burft their windows or walls outwards, pluck off the roofs, and pluck up the floors, by the fudden rarefaction of the air contained within fuch buildings; the outward preffure of the atmofphere being fuddenly taken off: So the ftopp'd bottle of air burfts under the exhaufted receiver of the air-pump.

FIG. II. is to reprefent the elevation of a water-fpout, wherein, I fuppofe P P P to be the cone, at firft a vacuum, till W W, the rifing column of water, has filled fo much of it. S S S S, the fpiral whirl of air furrounding the vacuum, and continued higher in a clofe column after the vacuum ends in the point P, till it reaches the cool region of the air. B B, the bufh defcribed by *Stuart*, furrounding the foot of the column of water.

Now, I fuppofe this whirl of air will, at firft, be as invifible as the air itfelf, though reaching, in reality, from the water, to the region of cool air, in which our low fummer thunder-clouds commonly float; but prefently it will become vifible at its extremities. *At its lower end*, by the agitation of the water, under the whirling part of the circle, between P and S, forming *Stuart*'s bufh, and by the fwelling and rifing of the water, in the beginning vacuum, which is, at firft, a fmall, low, broad cone, whofe top gradually rifes and fharpens, as the force of the whirl encreafes. *At its upper end* it becomes vifible, by the warm air brought up to the cooler region, where its moifture begins to be condenfed into thick vapour, by the cold, and is feen firft at A, the higheft part, which being

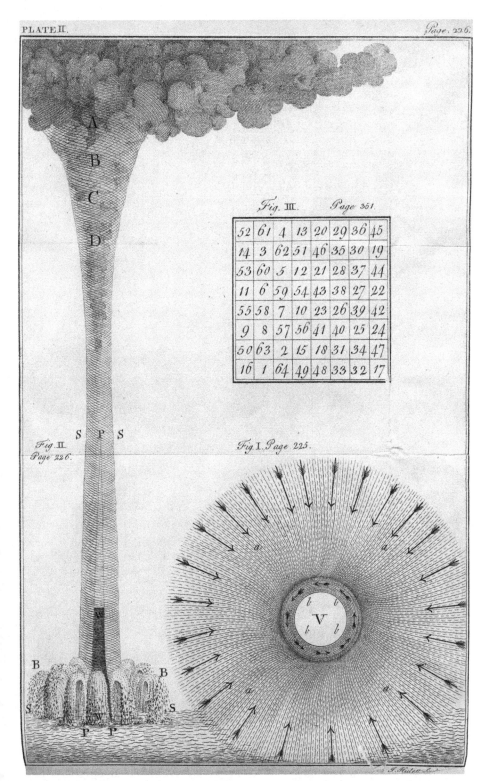

being now cooled, condenfes what rifes next at B, which condenfes that at C, and that condenfes what is rifing at D, the cold operating by the contact of the vapours fafter in a right line downwards, than the vapours themfelves can climb in a fpiral line upwards ; they climb, however, and as by continual addition they grow denfer, and, confequently, their centrifugal force greater, and being rifen above the concentrating currents that compofe the whirl, they fly off, fpread, and form a cloud.

It feems eafy to conceive, how, by this fucceffive condenfation from above, the fpout appears to drop or defcend from the cloud, though the materials of which it is compofed are all the while afcending.

The condenfation of the moifture contained in fo great a quantity of warm air as may be fuppofed to rife in a fhort time in this prodigioufly rapid whirl, is, perhaps, fufficient to form a great extent of cloud, though the fpout fhould be over land, as thofe at *Hatfield*; and if the land happens not to be very dufty, perhaps the lower part of the fpout will fcarce become vifible at all ; though the upper, or what is commonly called, the defcending part, be very diftinctly feen.

The fame may happen at fea, in cafe the whirl is not violent enough to make a high vacuum, and raife the column, &c. In fuch cafe, the upper part A B C D only will be vifible, and the bufh, perhaps, below.

But if the whirl be ftrong, and there be much duft on the land, and the column W W be raifed from the wa-

H h ter,

ter, then the lower part becomes vifible, and fometimes even united to the upper part. For the duft may be carried up in the fpiral whirl, till it reach the region where the vapour is condenfed, and rife with that even to the clouds : And the friction of the whirling air, on the fides of the column W W, may detach great quantities of its water, break it into drops, and carry them up in the fpiral whirl mixed with the air ; the heavier drops may, indeed, fly off, and fall, in a fhower, round the fpout; but much of it will be broken into vapour, yet vifible; and thus, in both cafes, by duft at land, and, by water at fea, the whole tube may be darkened and rendered vifible.

As the whirl weakens, the tube may (in appearance) feparate in the middle; the column of water fubfiding, and the fuperior condenfed part drawing up to the cloud. Yet ftill the tube, or whirl of air, may remain entire, the middle only becoming invifible, as not containing vifible matter.

Dr *Stuart* fays, ' It was obfervable of all the fpouts ' he faw, but more perceptible of the great one ; that, ' towards the end, it began to appear like a hollow canal, ' only black in the borders, but white in the middle; ' and though at firft it was altogether black and opaque, ' yet, now, one could very diftinctly perceive the fea- ' water to fly up along the middle of this canal, as fmoak ' up a chimney.'

And Dr *Mather*, defcribing a whirlwind, fays, ' a ' thick dark fmall cloud arofe, with a pillar of light in ' it,

' it, of about eight or ten feet diameter, and paſſed a-
' long the ground in a tract not wider than a ſtreet, hor-
' ribly tearing up trees by the roots, blowing them up in
' the air like feathers, and throwing up ſtones of great
' weight to a conſiderable height in the air, &c.'

Theſe accounts, the one of water-ſpouts, the cther of
a whirlwind, ſeem, in this particular, to agree; what one
Gentleman deſcribes as a tube, black in the borders, and
white in the middle, the other calls a black cloud, with
a pillar of light in it; the latter expreſſion has only a
little more of the *marvellous*, but the thing is the ſame;
and it ſeems not very difficult to underſtand. When Dr
Stuart's ſpouts were full charged, that is, when the whirling
pipe of air was filled between *a a a a* and *b b b b*, Fig. I.
with quantities of drops, and vapour torn off from the column W W, Fig. II. the whole was rendered ſo dark, as

that it could not be ſeen
through, nor the ſpiral a-
ſcending motion diſco-
vered; but when the
quantity aſcending leſſened, the pipe became more
tranſparent, and the a-
ſcending motion viſible.
For, by inſpection of this
figure in the margin, re-

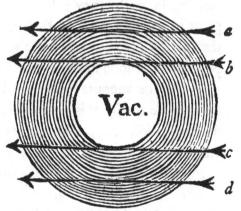

preſenting a ſection of our ſpout, with the vacuum in the
middle, it is plain that if we look at ſuch a hollow pipe in

H h 2　　　　the

the direction of the arrows, and suppose opaque particles to be equally mix'd in the space between the two circular lines, both the part between the arrows *a* and *b*, and that between the arrows *c* and *d*, will appear much darker than that between *b* and *c*, as there must be many more of those opaque particles in the line of vision across the sides, than across the middle. It is thus that a hair in a microscope evidently appears to be a pipe, the sides shewing darker than the middle. Dr *Mather*'s whirl was probably filled with dust, the sides were very dark, but the vacuum within rendering the middle more transparent, he calls it a pillar of light.

It was in this more transparent part, between *b* and *c*, that *Stuart* could see the spiral motion of the vapours, whose lines on the nearest and farthest side of the transparent part crossing each other, represented smoak ascending in a chimney ; for the quantity being still too great in the line of sight through the sides of the tube, the motion could not be discovered there, and so they represented the solid sides of the chimney.

When the vapours reach in the pipe from the clouds near to the earth, it is no wonder now to those who understand Electricity, that flashes of lightning should descend by the spout as in that at *Rome.*

But you object, If water may be thus carried into the clouds, why have we no salt rains ? The objection is strong and reasonable, and I know not whether I can answer it to your satisfaction. I never heard but of one salt
rain,

rain, and that was where a fpout paffed pretty near a
fhip, fo I fuppofe it to be only the drops thrown off
from the fpout, by the centrifugal force (as the birds were
at *Hatfield*) when they had been carried fo high as to be
above, or to be too ftrongly centrifugal for, the preffure
of the concurring winds furrounding it : And, indeed, I
believe there can be no other kind of falt rain ; for it has
pleafed the goodnefs of God fo to order it, that the particles of air will not attract the particles of falt, though
they ftrongly attract water.

Hence, though all metals, even gold, may be united
with air, and rendered volatile, falt remains fixt in the
fire, and no heat can force it up to any confiderable
height or oblige the air to hold it. Hence, when falt
rifes, as it will a little way, into air with water, there is
inftantly a feparation made ; the particles of water adhere
to the air, and the particles of falt fall down again, as if
repelled and forced off from the water by fome power in
the air ; or, as fome metals diffolved in a proper menftruum, will quit the folvent when other matter approaches,
and adhere to that, fo the water quits the falt, and embraces the air ; but air will not embrace the falt, and
quit the water, otherwife our rains would indeed be falt,
and every tree and plant on the face of the earth be deftroyed, with all the animals that depend on them for
fubfiftence. —— He who hath proportioned and given
proper qualities to all things, was not unmindful of this.
Let us adore Him with praife and thankfgiving !

By

By fome accounts of feamen, it feems the column of water W W, fometimes falls fuddenly; and if it be, as fome fay, fifteen or twenty yards diameter, it muft fall with great force, and they may well fear for their fhips. By one account, in the *Tranfactions*, of a fpout that fell at *Colne* in *Lancafhire*, one would think the column is fometimes lifted off from the water, and carried over land, and there let fall in a body.; but this, I fuppofe, happens rarely.

Stuart defcribes his fpouts as appearing no bigger than a maft, and fometimes lefs; but they were feen at a league and a half diftance.

I think I formerly read in *Dampier*, or fome other voyager, that a fpout, in its progreffive motion, went over a fhip becalm'd, on the coaft of *Guinea*, and firft threw her down on one fide, carrying away her fore-maft, then fuddenly whipp'd her up, and threw her down on the other fide, carrying away her mizen-maft, and the whole was over in an inftant. I fuppofe the firft mifchief was done by the fore-fide of the whirl, the latter by the hinder fide, their motion being contrary.

I fuppofe a whirlwind, or fpout, may be ftationary, when the concurring winds are equal; but if unequal, the whirl acquires a progreffive motion, in the direction of the ftrongeft preffure.

When the wind that gives the progreffive motion, becomes ftronger below than above, or above than below, the fpout will be bent, and the caufe ceafing, ftraiten again.

Your

Your Queries, towards the end of your paper, appear judicious, and worth confidering. At prefent I am not furnifhed with facts fufficient to make any pertinent anfwer to them ; and this paper has already a fufficient quantity of conjecture.

Your manner of accommodating the accounts to your hypothefis of defcending fpouts, is, I own, ingenious ; and perhaps that hypothefis may be true. I will confider it farther, but, as yet, I am not fatisfied with it, though hereafter I may be.

Here you have my method of accounting for the principal phænomena, which I fubmit to your candid examination.

And as I now feem to have almoft written a book, inftead of a letter, you will think it high time I fhould conclude ; which I beg leave to do, with affuring you that

I am, Sir, &c.

B. F.

LETTER XXI.

From Doctor *M——r*,

TO

ʙenjamin Franklin, Efq; at *Philadelphia.*

Read at the Royal Society, *June* 24, 1756.

SIR,   *New-Brunfwick, November* 11, 1752.

I AM favoured with your letter of the 2d inftant, and fhall, with pleafure, comply with your requeft, in defcribing (as well as my memory ferves me) the water-fpout I faw at *Antigua* ; and fhall think this, or any other fervice I can do, well repaid, if it contributes to your fatisfaction in fo curious a difquifition.

I had often feen water-fpouts at a diftance, and heard many ftrange ftories of them, but never knew any thing fatisfactory of their nature or caufe, until that which I faw at *Antigua* ; which convinced me that a water-fpout is a whirlwind, which becomes vifible in all its dimenfions by the water it carries up with it.

There appeared, not far from the mouth of the harbour of St *John*'s, two or three water-fpouts, one of which took

took its courfe up the harbour. Its progreffive motion was flow and unequal, not in a ftrait line, but, as it were, by jerks or ftarts. When juft by the wharff I ftood about 100 yards from it. There appeared in the water a circle of about twenty yards diameter, which, to me, had a dreadful, though pleafing appearance. The water in this circle was violently agitated, being whifked about, and carried up into the air with great rapidity and noife, and reflected a luftre, as if the fun fhined bright on that fpot, which was more confpicuous, as there appeared a dark circle around it. When it made the fhore, it carried up with the fame violence fhingles, ftaves *, large pieces of the roofs of houfes, &c. and one fmall wooden houfe it lifted entire from the foundation on which it ftood, and carried it to the diftance of fourteen feet, where it fettled without breaking or overfetting ; and, what is remarkable, though the whirlwind moved from Weft to Eaft, the houfe moved from Eaft to Weft. Two or three Negroes and a white woman, were killed by the fall of timber, which it carried up into the air, and dropt again. After paffing through the town, I believe it was foon diffipated ; for, except tearing a large limb from a tree, and part of the cover of a fugar-work near the town, I do not remember any farther damage done by it. I conclude, wifhing you fuccefs in your enquiry, and am, &c. W. M.

---

* I fuppofe fhingles, ftaves, timber, and other lumber, might be lying in quantities on the wharff, for fale, as brought from the Northern colonies. B. F.

I i LET-

LETTER XXII.

From Doctor ———, of *Boston*,

TO

BENJAMIN FRANKLIN, *Esq*; of *Philadelphia*.

SIR,   *Boston, May* 14, 1753.

Read at the Royal Society *July* 8, 1756.
I Received your letter of *April* laſt, and thank you for it. Several things in it make me at a loſs which ſide the truth lies on, and determine me to wait for farther evidence.

As to ſhooting ſtars, as they are called, I know very little, and hardly know what to ſay. I imagine them to be paſſes of electric fire from place to place in the atmoſphere, perhaps occaſioned by accidental preſſures of a non-electric circumambient fluid, and ſo by propulſion, or allicited by the circumſtance of a diſtant quantity *minus* electrified, which it ſhoots to ſupply, and becomes apparent by its contracted paſſage through a non-electric medium. Electric fire in our globe is always in action, ſometimes aſcending, deſcending, or paſſing from region to region

gion

gion. I fuppofe it avoids too dry air, and therefore we never fee thefe fhoots afcend. It always has freedom e-nough to pafs down unobferved, but, I imagine, not al-ways fo, to pafs to diftant climes and meridians lefs ftored with it.

The fhoots are fometimes all one way, which, in the laft cafe, they fhould be.

Poffibly there may be collections of particles in our at-mofphere, which gradually form, by attraction, either fimilar ones *per fe*, or diffimilar particles, by the interven-tion of others, But then, whether they fhoot or explode of themfelves, or by the approach of fome fuitable foreign collection, accidentally brought near by the ufual commo-tions and interchanges of our atmofphere, efpecially when the higher and lower regions intermix, before change of winds and weather, I leave.

I believe I have now faid enough of what I know no-thing about. If it fhould ferve for your amufement, or any way oblige you, it is all I aim at, and fhall, at your defire, be always ready to fay what I think, as I am fure of your candour.

I am, &c.

A Sub-

A subsequent PAPER *from the same.*

Read at the Royal
Society *July*
8, 1756.

SPOUTS have been generally believed ascents of water from below, to the region of the clouds, and whirlwinds the means of conveyance. The world has been very well satisfied with these opinions, and prejudiced with respect to any observations about them. Men of learning and capacity have had many opportunities in passing those regions where these phænomena were most frequent, but seem industriously to have declined any notice of them, unless to escape danger, as a matter of mere impertinence in a case so clear and certain as their nature and manner of operation are taken to be. Hence it has been very difficult to get any tolerable accounts of them. None but those they fell near can inform us any thing to be depended on; three or four such instances follow, where the vessels were so near, that their crews could not avoid knowing something remarkable with respect to the matters in question.

Capt. *John Wakefield*, junior, passing the *Streights* of *Gibraltar*, had one fall by the side of his ship; it came down of a sudden, as they think, and all agree the descent was certain.

Capt.

Capt. *Lang ftaff*, on a voyage to the *Weft-Indies*, had one come acrofs the ftern of his veffel, and paffed away from him. The water came down in fuch quantity that the prefent Capt. *Melling*, who was then a common failor at helm, fays it almoft drowned him, running into his mouth, nofe, ears, &c. and adds, that it tafted perfectly frefh.

One paffed by the fide of Capt. *Howland*'s fhip, fo near that it appeared pretty plain that the water defcended from firft to laft.

Mr *Robert Spring* was fo near one in the *Streights* of *Mallacca*, that he could perceive it to be a fmall very thick rain.

All thefe affure me, that there was no wind drawing towards them, nor have I found any others that have obferved fuch a wind.

It feems plain, by thefe few inftances, that whirlwinds do not always attend fpouts; and that the water really defcends in fome of them. But the following confideration, in confirmation of this opinion, may, perhaps, render it probable that all fpouts are defcents.

It feems unlikely that there fhould be two forts of fpouts, one afcending and the other defcending.

It has not yet been proved that any one fpout ever afcended. A fpecious appearance is all that can be produced in favour of this; and thofe who have been moft pofitive about it, were at more than a leagues diftance when they obferved, as *Stuart* and others, if I am not

miftaken

miftaken. However, I believe it impoffible to be certain whether water afcends or defcends at half the diftance.

It may not be amifs to confider the places where they happen moft. Thefe are fuch as are liable to calms from departing winds on both fides, as on the borders of the Æquinoctial trade, calms on the coaft of *Guinea*, in the *Streights* of *Malacca*, &c. places where the under region of the atmofphere is drawn off horizontally. I think they don't come where the calms are without departing winds; and I take the reafon to be, that fuch places, and places where winds blow towards one another, are liable to whirlwinds, or other afcents of the lower region, which I fuppofe contrary to fpouts. But the former are liable to defcents, which I take to be neceffary to their production. Agreable to this, it feems reafonable to believe, that any *Mediterranean* fea fhould be more fubject to fpouts than others. The fea ufually fo called is fo. The *Streights* of *Malacca* is. Some large gulphs may probably be fo, in fuitable latitudes; fo the *Red Sea*, &c. and all for this reafon, that the heated lands on each fide, draw off the under region of the air, and make the upper defcend, whence fudden and wonderful condenfations may take place, and make thefe defcents.

It feems to me, that the manner of their appearance and procedure, favour the notion of a defcent.

More or lefs of a cloud, as I am informed, always appears over the place firft; then a fpattering on the furface of the water below; and when this is advanced to a confiderable

fiderable degree, the fpout emerges from the cloud, and defcends, and that, if the caufes are fufficient, down to the places of fpattering, with a roaring in proportion to the quantity of the difcharge ; then it abates, or ftops, fometimes more gradually, fometimes more fuddenly.

I muft obferve a few things on thefe particulars, to fhew how I think they agree with my hypothefis.

The preceding cloud over the place fhews condenfation, and, confequently, tendency downwards, which therefore muft naturally prevent any afcent. Befides that, fo far as I can learn, a whirlwind never comes under a cloud, but in a clear fky.

The fpattering may be eafily conceived to be caufed by a ftream of drops, falling with great force on the place, imagining the fpout to begin fo, when a fudden and great condenfation happens in a contracted fpace, as the Ox Eye on the coaft of *Guinea.*

The fpout appearing to defcend from the cloud feems to be, by the ftream of nearly contiguous drops bringing the air into confent, fo as to carry down a quantity of the vapour of the cloud ; and the pointed appearance it makes may be from the defcending courfe being fwifteft in the middle, or center of the fpout. This naturally drawing the outer parts inward, and the center to a point ; and that will appear foremoft that moves fwifteft. The phænomenon of retiring and advancing, I think may be accounted for, by fuppofing the progreffive motion to exceed or not equal the confumption of the vapour by condendenfation. Or more plainly thus: The defcending vapour which forms the apparent fpout, if it be flow in its progrefs downwards, is condenfed as faft as it advances, and fo appears at a ftand; when it is condenfed fafter than it advances, it appears to retire; and *vice verfa*.

Its duration and manner of ending, are as the caufes, and may vary by feveral accidents.

The cloud itfelf may be fo circumftanced as to ftop it; as when, extending wide, it weighs down at a diftance round about, while a fmall circle at the fpout being exonerated by the difcharge, afcends and fhuts up the paffage. A new determination of wind may, perhaps, ftop it too. Places liable to thefe appearances are very liable to frequent and fudden alterations of it.

Such accidents as a clap of thunder, firing cannon, &c. may ftop them, and the reafon may be, that any fhock of this kind may occafion the particles that are near cohering, immediately to do fo; and then the whole, thus condenfed, falls at once (which is what I fuppofe is vulgarly called the breaking of the fpout) and in the interval, between this period and that of the next fet of particles being ready to unite, the fpout fhuts up. So that if this reafoning is juft, thefe phænomena agree with my hypothefis.

The ufual temper of the air, at the time of their appearance, if I have a right information, is for me too; it being then pretty cool for the feafon and climate; and this is worth remark, becaufe cool air is weighty, and will

not

not afcend ; befides, when the air grows cool, it fhews that the upper region defcends, and conveys this temper down ; and when the tempers are equal, no whirlwind can take place. But fpouts have been known, when the lower region has been really cold. *Gordon's* fpout in the *Downs* is an inftance of this—*(Vide Philofophical Tranf-actions*—) where the upper region was probably not at all cooler, if fo cold as the lower : It was a cold day in the month of *March* ; hail followed, but not fnow ; and it is obfervable, that not fo much as hail follows or accompanies them in moderate feafons or climes, when and where they are moft frequent. However, it is not improbable, that juft about the place of defcent may be cooler than the neighbouring parts, and fo favour the wonderful celerity of condenfation. But, after all, fhould we allow the under region to be ever fo much the hotteft, and a whirlwind to take place in it : Suppofe then the fea-water to afcend, it would certainly cool the fpout, and then, query, whether it would not very much, if not wholly, obftruct its progrefs.

It commonly rains when fpouts difappear, if it did not before, which it frequently does not, by the beft accounts I have had ; but the cloud encreafes much fafter after they difappear, and it foon rains. The firft fhews the fpout to be a contracted rain, inftead of the diffufed one that follows ; and the latter that the cloud was not formed by afcending water, for then it would have ceafed growing when the fpout vanifhed.

K k

However

However, it feems that fpouts have fometimes appeared after it began to rain; but this is one way a proof of my hypothefis, *viz.* as whirlwinds don't come under a cloud.

I forgot to mention, that the increafe of cloud, while the fpout fubfifts, is no argument of an afcent of water, by the fpout. Since thunder-clouds fometimes encreafe greatly while it rains very hard.

Divers effects of fpouts feem not fo well accounted for any other way as by defcent.

The bufh round the feet of them feems to be a great fpray of water made by the violence of defcent, like that in great falls of water from high precipices.

The great roar, like fome vaft inland falls, is fo different from the roar of whirlwinds, by all accounts, as to be no ways compatible.

The throwing things from it with great force, inftead of carrying them up into the air, is another difference.

There feems fome probability that the failors traditionary belief that fpouts may break in their decks, and fo deftroy veffels, might originate from fome facts of that fort in former times. This danger is apparent on my hypothefis, but it feems not fo on the other : And my reafon for it is, that the whole column of a fpout from the fea to the clouds, cannot, in a natural way, even upon the largeft fuppofition, fupport more than about three feet water, and from truly fuppofeable caufes, not above one foot, as may appear more plainly by and by. Suppofing posing now the largeft of thefe quantities to rife, it muft be diffeminated into drops, from the furface of the fea to the region of the clouds, or higher; for this reafon it is quite unlikely to be collected into maffes, or a body, upon its falling; but would defcend in progreffion according to the feveral degrees of altitude the different portions had arrived at when it received this new determination.

Now that there cannot more rife upon the common hypothefis than I have mentioned, may appear probable, if we attend to the only efficient caufe in fuppofed a-fcending fpouts, *viz.* whirlwinds.

We know that the rarefaction of the lower, and the condenfation of the upper region of air, are the only natural caufes of whirlwinds. Let us then fuppofe the former as hot as their greateft fummer heat in *England,* and the latter as cold as the extent of their winter. Thefe extremes have been found there to alter the weight of the air one-tenth, which is equal to a little more than three feet water. Were this cafe poffible, and a whirlwind take place in it, it might act with a force equal to the mentioned difference. But as this is the whole ftrength, fo much water could not rife; therefore to allow it due motion upwards, we muft abate, at leaft, one-fourth part, perhaps more, to give it fuch a fwift afcenfion as fome think ufual. But here feveral difficulties occur, at leaft they are fo to me. As, whether this quantity would ren-der the fpout opaque? Since it is plain that in drops it could not do fo. How, or by what means it may be re-

duced

duced fmall enough ? Or, if the water be not reduced
into vapour, what will fufpend it in the region of the
clouds when exonerated there ? And, if vapourized while
afcending, how it can be dangerous by what they call
the breaking ? For it is difficult to conceive how a con-
denfative power fhould inftantaneoufly take place of a ra-
rifying and diffeminating one.

The fudden fall of the fpout, or, rather, the fudden
ceafing of it, I accounted for, in my way, before. But it
feems neceffary to mention fomething I then forgot.
Should it be faid to do fo, (*i. e.*) to fall, becaufe all the
lower rarified air is afcended, whence the whirlwind muft
ceafe, and its burden drop ; I cannot agree to this, unlefs
the air be obferved on a fudden to have grown much
colder, which I can't learn has been the cafe. Or fhould
it be fuppofed that the fpout was, on a fudden, obftructed at the top, and this the caufe of the fall, however
plaufible this might appear, yet no more water would
fall than what was at the fame time contained in the
column, which is often, by many and fatisfactory accounts to me, again far from being the cafe.

We are, I think, fufficiently affured, that not only
tons, but fcores or hundreds of tons defcend in one fpout.
Scores of tons more than can be contained in the trunk
of it, fhould we fuppofe water to afcend.

But, after all, it don't appear that the above-mentioned different degrees of heat and cold concur in any region
where fpouts ufually happen, nor, indeed, in any other.

Ob-

Observations on the METEOROLOGICAL PAPER ;
by a Gentleman in Connecticut.

Read at the Royal
Society, *Nov.* 4,
1756.
" **A**IR and water mutually attract each other, (faith Mr *F.*) hence water will diffolve in air, as falt in water." I think that he hath demonftrated, that the fupporting of falt in water is not owing to its fuperficies being increafed, becaufe " the fpecific gravity of falt is not altered by dividing of it, any more than that of lead, fixteen bullets of which, of an ounce each, weigh as much in water as one of a pound." But yet, when this came to be applied to the fupporting of water in air, I found an objection rifing in my mind.

In the firft place, I have always been loth to feek for any new hypothefis, or particular law of nature, to account for any thing that may be accounted for from the known, general, and univerfal law of nature ; it being an argument of the infinite wifdom of the Author of the World, to effect fo many things by one general law. Now I had thought that the rifing and fupport of water in air, might be accounted for from the general law of gravitation, by only fuppofing the fpaces occupied by the fame quantity of water increafed.

And

And, wirh refpect to the lead, I queried thus in my own mind ; whether if the fuperficies of a bullet of lead fhould be increafed four or five fold by an internal vacuity, it would weigh the fame in water as before. I mean; if a pound of lead fhould be formed into a hollow globe, empty within, whofe fuperficies fhould be four or five times as big as that of the fame lead when a folid lump, it would weigh as much in water as before. I fuppofed it would not. If this concavity was filled with water, perhaps it might ; if with air, it would weigh at leaft as much lefs, as this difference between the weight of that included air, and that of water.

Now although this would do nothing to account for the diffolution of falt in water, the fmalleft lumps of falt being no more hollow fpheres, or any thing of the like nature than the greateft ; yet, perhaps, it might account for water's rifing and being fupported in air. For you know that fuch hollow globules, or bubbles, abound upon the furface of the water, which even by the breath of our mouths, we can caufe to quit the water, and rife in the air.

Thefe bubbles I ufed to fuppofe to be coats of water, containing within them air rarified and expanded with fire, and that, therefore, the more friction and dafhing there is upon the furface of the waters, and the more heat and fire, the more they abound.

And I ufed to think, that although water be fpecifically heavier than air, yet fuch a bubble, filled only with fire

and

and very rarified air, may be lighter than a quantity of common air, of the fame cubical dimenfions, and, therefore, afcend ; for the rarified air inclofed, may more fall fhort of the fame bulk of common air, in weight, than the watery coat exceeds a like bulk of common air in gravity.

This was the objection in my mind, though, I muft confefs, I know not how to account for the watery coat's encompaffing the air, as above-mentioned, without allowing the attraction between air and water, which the Gentleman fuppofes ; fo that I don't know but that this objection examined by that fagacious Genius, will be an additional confirmation of the hypothefis.

The Gentleman obferves, " That a certain quantity of moifture fhould be every moment difcharged and taken away from the lungs ;" and hence accounts for the fuffocating nature of fnuffs of candles, as impregnating the air with greafe, between which and water there is a natural repellency ; and of air that hath been frequently breathed in, which is overloaded with water, and, for that reafon, can take no more air. Perhaps the fame obfervation will account for the fuffocating nature of damps in wells.

But then if the air can fupport and take off but fuch a proportion of water, and it is neceffary that water be fo taken off from the lungs, I queried with myfelf how it is we can breathe in an air full of vapours, fo full as that they continually precipitated. Don't we fee the air o-

ver-

verloaded, and cafting forth water plentifully when there is no fuffocation ?

The Gentleman again obferves, " That the air under " the Equator, and between the Tropics, being conftantly " heated and rarified by the fun, rifes ; its place is fup- " plied by air from Northern and Southern latitudes, " which coming from parts where the air and earth had " lefs motion, and not fuddenly acquiring the quicker mo- " tion of the equatorial earth, appears an Eaft wind blow- " ing Weftward ; the earth moving from Weft to Eaft, " and flipping under the air."

In reading this, two objections occurred to my mind :

Firft, That it is faid, the trade-wind doth not blow in the forenoon, but only in the afternoon.

Secondly, That either the motion of the Northern and Southern air towards the Equator is fo flow, as to acquire almoft the fame motion as the equatorial air when it ar- rives there, fo that there will be no fenfible difference ; or elfe the motion of the Northern and Southern air to- wards the Equator, is quicker, and muft be fenfible ; and then the trade-wind muft appear either as a South-Eaft or North-Eaft wind : South of the Equator, a South-Eaft wind ; North of the Equator a North-Eaft. For the ap- parent wind muft be compounded of this motion from North to South, or *vice verfa* ; and of the difference be- tween its motion from Weft to Eaft, and that of the equa- torial air.

Ob-

Obfervations in Anfwer to the foregoing ; by B. F.

Read at the Royal-
Society, *Nov.* 4,
1756. 1ft. THE fuppofing a mutual attraction between the particles of water and air, is not
introducing a new law of nature ; fuch attractions taking
place in many other known inftances.

2dly. Water is fpecifically 850 times heavier than air.
To render a bubble of water, then, fpecifically lighter
than air, it feems to me that it muft take up more than
850 times the fpace it did before it formed the bubble ;
and within the bubble fhould be either a vacuum or air
rarified more than 850 times. If a vacuum, would not
the bubble be immediately crufh'd by the weight of the
atmofphere ? And no heat, we know of, will rarify air
any thing near fo much ; much lefs the common heat of
the fun, or that of friction by the dafhing on the furface of the water. Befides, water agitated ever fo violently, produces no heat, as has been found by accurate
experiments.

3dly. A hollow fphere of lead has a firmnefs and con-
fiftency in it, that a hollow fphere or bubble of fluid un-
frozen water cannot be fuppofed to have. The lead may
fupport the preffure of the water it is immerged in, but

L l the

the bubble could not fupport the preffure of the air, if empty within.

4thly. Was ever a vifible bubble feen to rife in air? I have made many, when a boy, with foap-fuds and a tobacco-pipe; but they all defcended when loofe from the pipe, though flowly, the air impeding their motion. They may, indeed, be forced up by a wind from below, but do not rife of themfelves, though filled with warm breath.

5thly. The objection relating to our breathing moift air, feems weighty, and muft be farther confidered. The air that has been breathed, has, doubtlefs, acquired an addition of the perfpirable matter which nature intends to free the body from, and which would be pernicious if retained and returned into the blood; fuch air then may become unfit for refpiration, as well for that reafon, as on account of its moifture. Yet I fhould be glad to learn, by fome accurate experiment, whether a draft of air, two or three times infpired, and expired, perhaps in a bladder, has, or has not, acquired more moifture than our common air in the dampeft weather. As to the pre-cipitation of water in the air we breathe, perhaps it is not always a mark of that air's being overloaded. In the region of the clouds, indeed, the air muft be overload-ed if it lets fall its water in drops, which we call rain; but thofe drops may fall through a dryer air near the earth; and accordingly we find that the hygrofcope fome-times fhews a lefs degree of moifture, during a fhower,

than

than at other times when it does not rain at all. The dewy dampness that settles on the insides of our walls and wainscots, seems more certainly to denote an air overloaded with moisture ; and yet this is no sure sign : For, after a long continued cold season, if the air grows suddenly warm, the walls, &c. continuing longer their coldness, will, for some time, condense the moisture of such air, till they grow equally warm, and then they condense no more, though the air is not become dryer. And, on the other hand, after a warm spell, if the air grows cold, though moister than before, the dew is not so apt to gather on the walls. A tankard of cold water will, in a hot and dry summer's day, collect a dew on its outside; a tankard of hot water will collect none in the moistest weather.

6thly. It is, I think, a mistake that the trade-winds blow only in the afternoon. They blow all day and all night, and all the year round, except in some particular places. The southerly sea-breezes on your coasts, indeed, blow chiefly in the afternoon. In the very long run, from the West side of *America*, to *Guam*, among the *Phillippine Islands*, ships seldom have occasion to hand their sails, so equal and steady is the gale, and yet they make it in about 60 days, which could not be, if the wind blew only in the afternoon.

7thly. That really is, which the Gentleman justly supposes ought to be on my hypothesis. In sailing

L l 2

South-

Southward, when you firft enter the trade-wind, you find it North-Eaft, or thereabouts, and it gradually grows more Eaft as you approach the line. The fame obferva-tion is made of its changing from South-Eaft to Eaft gra-dually, as you come from the Southern latitudes to the equator.

Obfervations on the METEOROLOGICAL PAPER; *fent by a Gentleman in* New-York, *to* B. F.

Read at the Royal-
Society, *Nov.* 4,
1756.

THAT power by which the air expands itfelf, you attribute to a mutual repelling power in the particles which compofe the air, by which they are fepa-rated from each other with fome degree of force : Now this force, on this fuppofition, muft not only act when the particles are in mutual contact, but likewife when they are at fome diftance from each other. How can two bodies, whether they be great or fmall, act at any diftance, whether that diftance be fmall or great, without fomething intermediate on which they act ? For if any body act on another, at any diftance from it, however fmall that diftance be, without fome medium to conti-nue the action, it muft act where it is not, which to me feems abfurd.

It

It feems to me, for the fame reafon, equally abfurd to give a mutual attractive power between any other particles fuppofed to be at a diftance from each other, without any thing intermediate to continue their mutual action. I can neither attract nor repel any thing at a diftance, without fomething between my hand and that thing, like a ftring, or a ftick; nor can I conceive any mutual action without fome middle thing, when the action is continued to fome diftance.

The encreafe of the furface of any body leffens its weight, both in air, and water, or any other fluid, as appears by the flow defcent of leaf-gold in the air.

The obfervation of the different denfity of the upper and lower air, from heat and cold, is good, and I do not remember it is taken notice of by others; the confequences alfo are well drawn; but as to winds, they feem principally to arife from fome other caufe. Winds generally blow from fome large tracts of land, and from mountains. Where I live, on the North fide of the mountains, we frequently have a ftrong Southerly wind, when they have as ftrong a Northerly wind, or calm, on the other fide of thefe mountains. The continual paffing of veffels on *Hudfon's River*, through thefe mountains, give frequent opportunities of obferving this.

In the fpring of the year the fea-wind (by a piercing cold) is always more uneafy to me, accuftomed to winds which pafs over a tract of land, than the North-Weft wind.

You

You have received the common notion of water-spouts, which, from my own ocular obfervation, I am perfuaded is a falfe conception. In a voyage to the *Weft-Indies* I had an opportunity of obferving many water-fpouts. One of them paffed nearer than thirty or forty yards to the veffel I was in, which I viewed with a good deal of attention; and though it be now forty years fince I faw it, it made fo ftrong an impreffion on me, that I very diftinctly remember it. Thefe water-fpouts were in the calm latitudes, that is, between the trade and the variable winds, in the month of *July*. That fpout which paffed fo near us, was an inverted cone, with the *tip* or *apex* towards the fea, and reached within about eight feet of the furface of the fea, its bafis in a large black cloud. We were entirely becalmed. It paffed flowly by the veffel. I could plainly obferve that a violent ftream of wind iffued from the fpout, which made a hollow of about fix feet diameter in the furface of the water, and raifed the water in a circular uneven ring round the hollow, in the fame manner that a ftrong blaft from a pair of bellows would do when the pipe is placed perpendicular to the furface of the water; and we plainly heard the fame hiffing noife which fuch a blaft of wind muft produce on the water. I am very fui there was nothing like the fucking of water from the fea into the fpout, unlefs the fpray which was raifed in a ring to a fmall height, could be miftaken for a raifing of water. I could plainly diftinguifh a diftance of about eight feet between the fea and

the

the tip of the cone, in which nothing interrupted the fight, which muſt have been, had the water been raiſed from the ſea.

In the ſame voyage I ſaw ſeveral other ſpouts at a greater diſtance, but none of them whoſe tip of the cone came ſo near the ſurface of the water. In ſome of them the axis of the cone was conſiderably inclined from the perpendicular, but in none of them was there the leaſt appearance of ſucking up of water. Others of them were bent or arched. I believe that a ſtream of wind iſſued from all of them, and it is from this ſtream of wind that veſſels are often overſet, or founder at ſea ſuddenly. I have heard of veſſels being overſet when it was perfectly calm, the inſtant before the ſtream of wind ſtruck them, and immediately after they were overſet ; which could not otherwiſe be but by ſuch a ſtream of wind from a cloud.

That wind is generated in clouds will not admit of a diſpute. Now if ſuch wind be generated within the body of the cloud, and iſſue in one particular place, while it finds no paſſage in the other parts of the cloud, I think it may not be difficult to account for all the appearances in water-ſpouts ; and from hence the reaſon of breaking thoſe ſpouts, by firing a cannon-ball through them, as thereby a horizontal vent is given to the wind. When the wind is ſpent, which dilated the cloud, or the fermentation ceaſes, which generates the air and wind, the clouds may deſcend in a prodigious fall of water or rain.

A re-

A remarkable inteftine motion, like a violent fermentation, is very obfervable in the cloud from whence the fpout
iffues. No falt water, I am perfuaded, was ever obferved
to fall from the clouds, which muft certainly have happened if fea-water had been raifed by a fpout.

ANSWER *to the foregoing Obfervations*; *by* B. F.

Read at the Royal
Society, *Nov.*
4, 1756.
I Agree with you, that it feems abfurd
to fuppofe that a body can act where it
is not. I have no idea of bodies at a
diftance attracting or repelling one another without the
affiftance of fome medium, though I know not what that
medium is, or how it operates. When I fpeak of attraction or repulfion, I make ufe of thofe words for want
of others more proper, and intend only to exprefs *effects*
which I fee, and not *caufes* of which I am ignorant. When
I prefs a blown bladder between my knees, I find I cannot bring its fides together, but my knees feel a fpringy
matter, pufhing them back to a greater diftance, or repelling them. I conclude that the air it contains is the
caufe. And when I operate on the air, and find I cannot
by preffure force its particles into contact, but they ftill
fpring back againft the preffure, I conceive there muft be
fome medium between its particles that prevents their
clofing, though I cannot tell what it is.—And if I were

ac-

acquainted with that medium, and found its particles to approach and recede from each other, according to the preffure they fuffered, I fhould imagine there muft be fome finer medium between them, by which thefe operations were performed.

I allow that increafe of the furface of a body may occafion it to defcend flower in air, water, or any other fluid ; but do not conceive, therefore, that it leffens its weight. Where the increafed furface is fo difpofed as that in its falling a greater quantity of the fluid it finks in muft be moved out of its way, a greater time is required for fuch removal. Four fquare feet of fheet lead finking in water *broadways,* cannot defcend near fo faft as it would *edgeways,* yet its weight in the hydroftatic ballance would, I imagine, be the fame, whether fufpended by the middle or by the corner.

I make no doubt but that ridges of high mountains do often interrupt, ftop, reverberate, or turn the winds that blow againft them, according to the different degrees of ftrength of the winds, and the angles of incidence. I fuppofe, too, that the cold upper parts of mountains may condenfe the warmer air that comes near them, and fo by making it fpecifically heavier, caufe it to defcend on one or both fides of the ridge into the warmer valleys, which will feem a wind blowing from the mountain.

Damp winds, though not colder by the thermometer, give a more uneafy fenfation of cold than dry ones ; becaufe (to fpeak like an Electrician) they *conduct* better;

M m

that is, are better fitted to convey away the heat from our bodies. The body cannot feel *without* itſelf; our ſenſation of cold is not in the air *without* the body, but in thoſe parts of the body which have been deprived of their heat by the air. My deſk, and its lock, are, I ſuppoſe, of the ſame temperament when they have been long ex-poſed to the ſame air ; but now if I lay my hand on the wood, it does not ſeem ſo cold to me as the lock ; be-cauſe (as I imagine) wood is not ſo good a conductor, to receive and convey away the heat from my ſkin, and the adjacent fleſh, as metal is. Take a piece of wood, of the ſize and ſhape of a dollar, between the thumb and finger of one hand, and a dollar, in like manner, with the other hand ; place the edges of both, at the ſame time, in the flame of a candle ; and though the edge of the wooden piece takes flame, and the metal piece does not, yet you will be obliged to drop the latter before the former, it conducting the heat more ſuddenly to your fingers. Thus we can, without pain, handle glaſs and china cups filled with hot liquors, as tea, *&c.* but not ſilver ones. A ſilver tea-pot muſt have a wooden handle. Perhaps it is for the ſame reaſon that woollen garments keep the body warmer than linnen ones equally thick ; woollen keep-ing the natural heat in, or, in other words, not conducting it out to air.

In regard to water-ſpouts, having, in a long letter to a Gentleman of the ſame ſentiment with you as to their direction, ſaid all that I have to ſay in ſupport of my o-pinion ;

pinion; I need not repeat the arguments therein contained, as I intend to fend you a copy of it by fome other opportunity, for your perufal. I imagine you will find all the appearances you faw, accounted for by my hypothefis. I thank you for communicating the account of them. At prefent I would only fay, that the opinion of winds being generated in clouds by fermentation, is new to me, and I am unacquainted with the facts on which it is founded. I likewife find it difficult to conceive of winds confined in the body of clouds, which I imagine have little more folidity than the fogs on the earth's furface. The objection from the frefhnefs of rain-water is a ftrong one, but I think I have anfwered it in the letter abovementioned, to which I muft beg leave, at prefent, to refer you.

LET-

LETTER XXIII.

FROM

BENJ. FRANKLIN, *Efq*; of *Philadelphia*.

To C. C. Efq; at *New-York*.

S I R,   *Philadelphia, April* 23, 1752.

Read at the Royal
Society *Nov.*
11, 1756.
IN confidering your favour of the 16th paft, I recollected my having wrote you anfwers to fome queries concerning the difference between electrics *per fe*, and non-electrics, and the effects of air in electrical experiments, which, I apprehend, you may not have received. The date I have forgot.

We have been ufed to call thofe bodies electrics *per fe*, which would not conduct the electric fluid : We once imagined that only fuch bodies contained that fluid ; afterwards that they had none of it, and only educ'd it from other bodies : But further experiments fhewed our miftakes. It is to be found in all matter we know of; and the diftinctions of electrics *per fe*, and non-electrics, fhould now be dropt as improper, and that of *conductors* and *non-conductors* affumed in its place, as I mentioned in thofe anfwers.

I do

I do not remember any experiment by which it appeared that high rectified fpirit will not conduct ; perhaps you have made fuch. This I know, that wax, rofin, brimftone, and even glafs, commonly reputed electrics *per fe*, will, when in a fluid ftate, conduct pretty well. Glafs will do it when only red hot. So that my former pofition, that only metals and water were conductors, and other bodies more or lefs fuch, as they partook of metal or moifture, was too general.

Your conception of the electric fluid, that it is incomparably more fubtle than air, is undoubtedly juft. It pervades denfe matter with the greateft eafe ; but it does not feem to mix or incorporate willingly with meer air, as it does with other matter. It will not quit common matter to join with air. Air obftructs, in fome degree, its motion. An electric atmofphere cannot be communicated at fo great a diftance, through intervening air, by far, as through a vacuum.—Who knows then, but there may be,. as the Antients thought, a region of this fire above our atmofphere, prevented by our air, and its own too great diftance for attraction, from joining our earth ? Perhaps where the atmofphere is rareft, this fluid may be denfeft, and nearer the earth where the atmofphere grows denfer, this fluid may be rarer ; yet fome of it be low enough to attach itfelf to our higheft clouds, and thence they becoming electrified, may be attracted by, and defcend towards the earth, and difcharge their watry contents, together with that etherial fire. Perhaps the Auroræ Boreales,

ales are currents of this fluid in its own region, above our atmofphere, becoming from their motion vifible. There is no end to conjectures. As yet we are but novices in this branch of natural knowledge.

You mention feveral differences of falts in electrical experiments? Were they all equally dry? Salt is apt to acquire moifture from a moift air, and fome forts more than others. When perfectly dried by lying before a fire, or on a ftove, none that I have tried will conduct any better than fo much glafs.

New flannel, if dry and warm, will draw the electric fluid from non-electrics, as well as that which has been worn.

I wifh you had the convenience of trying the experiments you feem to have fuch expectations from, upon various kinds of fpirits, falts, earth, &c. Frequently, in a variety of experiments, though we mifs what we expected to find, yet fomething valuable turns out, fomething furprifing, and inftructing, though unthought of.

I thank you for communicating the illuftration of the theorem concerning light. It is very curious. But I muft own I am much in the *dark* about *light*. I am not fatisfied with the doctrine that fuppofes particles of matter called light, continually driven off from the fun's furface, with a fwiftnefs fo prodigious! Muft not the fmalleft particle conceivable, have with fuch a motion, a force exceeding that of a twenty-four pounder, difcharged from a cannon? Muft not the fun diminifh exceedingly

by

by such a waste of matter; and the planets, instead of drawing nearer to him, as some have feared, recede to greater distances through the lessened attraction. Yet these particles, with this amazing motion, will not drive before them, or remove, the least and lightest dust they meet with : And the sun, for aught we know, continues of his antient dimensions, and his attendants move in their antient orbits.

May not all the phænomena of light be more conveniently solved, by supposing universal space filled with a subtle elastic fluid, which, when at rest, is not visible, but whose vibrations affect that fine sense in the eye, as those of air do the grosser organs of the ear ? We do not, in the case of sound, imagine that any sonorous particles are thrown off from a bell, for instance, and fly in strait lines to the ear; why must we believe that luminous particles leave the sun and proceed to the eye? Some diamonds, if rubbed, shine in the dark, without losing any part of their matter. I can make an electrical spark as big as the flame of a candle, much brighter, and, therefore, visible further; yet this is without fuel; and, I am persuaded, no part of the electric fluid flies off in such case, to distant places, but all goes directly, and is to be found in the place to which I destine it. May not different degrees of the vibration of the above-mentioned universal medium, occasion the appearances of different colours? I think the electric fluid is always the same; yet I find that weaker and stronger sparks differ in apparent

rent

rent colour, fome white, blue, purple, red ; the ftrongeft, white ; weak ones red. Thus different degrees of vibration given to the air, produce the feven different founds in mufic, analagous to the feven colours, yet the medium, air, is the fame.

If the fun is not wafted by expence of light, I can eafily conceive that he fhall otherwife always retain the fame quantity of matter ; though we . fhould fuppofe him made of fulphur conftantly flaming. The action of fire only *feparates* the particles of matter, it does not *annihilate* them. Water, by heat raifed in vapour, returns to the earth in rain ; and if we could collect all the particles of burning matter that go off in fmoak, perhaps they might, with the afhes, weigh as much as the body before it was fired : And if we could put them into the fame pofition with regard to each other, the mafs would be the fame as before, and might be burnt over again. The chymifts have analyfed fulphur, and find it compofed, in certain proportions, of oil, falt, and earth ; and having, by the a- nalyfis, difcovered thofe proportions, they can, of thofe in- gredients, make fulphur. So we have only to fuppofe, that the parts of the fun's fulphur, feparated by fire, rife into his atmofphere, and there being freed from the im- mediate action of the fire, they collect into cloudy maffes, and growing, by degrees, too heavy to be longer fupported, they defcend to the fun, and are burnt over again. Hence the fpots appearing on his face, which are obferved to diminifh

minifh daily in fize, their confuming edges being of particular brightnefs.

It is well we are not, as poor *Galileo* was, fubject to the Inquifition for *Philosophical Herefy.* My whifpers againft the orthodox doctrine, in private letters, would be dangerous ; but your writing and printing would be highly criminal. As it is, you muft expect fome cenfure, but one Heretic will furely excufe another.

I am heartily glad to hear more inftances of the fuccefs of the Poke-Weed, in the cure of that horrible evil to the human body, a Cancer. You will deferve highly of mankind for the communication. But I find in *Bofton* they are at a lofs to know the right plant, fome afferting it is what they call *Mechoachan,* others other things. In one of their late papers it is publickly requefted that a perfect defcription may be given of the plant, its places of growth, &c. I have miflaid the paper, or would fend it to you. I thought you had defcribed it pretty fully.

I am, Sir, &c.

B. F.

Extracts from DAMPIER's *Voyages, rela-ting to* WATER-SPOUTS.

Read at the Royal Society, *Dec.* 16, 1756. " A Spout is a small ragged piece, or part of a cloud, hanging down about a yard seemingly, from the blackest part thereof. Commonly it hangs down sloping from thence, or sometimes appearing with a small bending, or elbow, in the middle. I never saw any hang perpendicularly down. It is small at the lower end, seeming no bigger than one's arm, but still fuller towards the cloud from whence it proceeds.

When the surface of the sea begins to work, you shall see the water for about one hundred paces in circumference foam and move gently round, till the whirling motion increases; and then it flies upward in a pillar, about one hundred paces in compass at the bottom, but gradually lessening upwards, to the smallness of the spout itselfs through which the rising sea-water seems to be conveyed into the clouds. This visibly appears by the clouds increasing in bulk and blackness. Then you shall presently see the cloud drive along, though before it seemed to be without any motion. The spout also keeping the same course with the cloud, and still sucking up the water as it goes along, and they make a wind as they go. Thus it

con-

continues for half an hour, more or less, until the fucking is spent, and then breaking off, all the water which was below the spout, or pendulous piece of cloud, falls down again into the sea, making a great noise with its falling and clashing motion in the sea.

It is very dangerous for a ship to be under a spout when it breaks; therefore we always endeavour to shun it, by keeping at a distance, if possibly we can. But for want of wind to carry us away, we are often in great fear and danger, for it is usually calm when spouts are at work, except only just where they are. Therefore men at sea, when they see a spout a coming, and know not how to avoid it, do sometimes fire shot out of their great guns into it, to give it air or vent, that so it may break; but I did never hear that it proved to be of any benefit.

And now we are on this subject, I think it not amiss to give you an account of an accident that happened to a ship once on the coast of *Guinea*, some time in or about the year 1674. One Capt. *Records* of *London*, bound for the coast of *Guinea*, in a ship of three hundred tons, and sixteen guns, called the *Blessing*, when he came into latitude seven or eight degrees North, he saw several spouts, one of which came directly towards the ship, and he having no wind to get out of the way of the spout, made ready to receive it by furling the sails. It came on very swift, and broke a little before it reached the ship, making a great noise, and raising the sea round it, as if a great house, or some such thing, had been cast into the sea.

       The

The fury of the wind still lasted, and took the ship on the star-board-bow with such violence, that it snapt off the boltsprit and foremast both at once, and blew the ship all along, ready to overset it; but the ship did presently right again, and the wind whirling round, took the ship a second time with the like fury as before, but on the contrary side, and was again like to overset her the other way: The mizen-mast felt the fury of this second blast, and was snapt short off, as the foremast and boltsprit had been before. The main-mast and main-top-mast received no damage, for the fury of the wind (which was presently over) did not reach them. Three men were in the fore-top when the foremast broke, and one on the boltsprit, and fell with them into the sea, but all of them were saved. I had this relation from Mr *John Canby*, who was then quarter-master and steward of her; one *Abraham Wise* was chief-mate, and *Leonard Jefferies* second-mate.

We are usually much afraid of them, yet this was the only damage that I ever heard done by them. They seem terrible enough, the rather because they come upon you while you lie becalmed, like a log in the sea, and cannot get out of their Way. But though I have seen and been beset by them often, yet the fright was always the greatest of the harm."——*Dampier*, Vol. I. page 451.

An

An Account of a SPOUT *on the Coast of* New-Guinea. *From the same.*

" WE had fair clear weather, and a fine moderate gale from South-East to East by North ; but at day-break the clouds began to fly, and it lightened very much in the East North-East. At sun rising the sky looked very red in the East near the horizon ; and there were many black clouds both to the South and North of it. About a quarter of an hour after the sun was up, there was a squall to the windward of us, when, on a sudden, one of our men on the fore-castle, called out that he saw something a-stern, but could not tell what. I looked out for it, and immediately saw a spout beginning to work within a quarter of a mile of us, exactly in the wind: We presently put right before it. It came very swiftly, whirling the water up in a pillar, about six or seven yards high. As yet I could not see any pendulous cloud from whence it might come ; and was in hopes it would soon lose its force   In four or five minutes time it came within a cable's length of us, and passed away to leeward ; and then I saw a long pale stream coming down to the whirling water. This stream was about the bigness of a rainbow. The upper end seemed vastly high, not deſcending from any dark cloud, and, therefore, the more ſtrange to me, I never having ſeen the like before. It paſt about a mile to the leeward of us, and then broke. This was but a ſmall ſpout, and not ſtrong nor *laſting ; yet I perceived much wind in it as it paſſed by us.   Vol. III. page 223.

Account of another SPOUT.   *From the ſame.*

" WE ſaw a ſpout but a ſmall diſtance from us ; it fell down out of a black cloud that yielded great ſtore of rain, thunder, and lightning.   This cloud hovered to the Southward of us for the ſpace of three hours, and then drew to the Weſtward a great pace, at which time it was that we ſaw the ſpout, which hung faſt to the cloud till it broke, and then the cloud whirled about to the South-Eaſt, then to the North-Eaſt, where meeting with an iſland, it ſpent itſelf, and ſo diſperſed ; and immediately we had a little of the tail of it, having had none before."   Vol III. page 182.

---

* Probably if it had been laſting, a cloud would have been formed a-bove it.   Theſe extracts from *Dampier*, ſeem, in different inſtances, to favour both opinions, and, therefore, are inſerted entire, for the Reader's conſideration.

Extract

Extract of a Letter from J. B. *Esq; in* Boston, *to* B. F. *concerning the Light in Sea-Water.*

November 12, 1753.

Read at the Royal
Society, *Dec.*
16, 1756.

WHEN I was at the Eastward, I had an opportunity of observing the luminous appearance of the sea when disturbed : At the head and stern of the vessel, when under way, it appeared very bright. The best opportunity I had to observe it, was in a boat, in company with several Gentlemen going from *Portsmouth*, about three miles, to our vessel lying at the mouth of *Piscataqua* River. Soon after we set off (it being in the evening) we observed a luminous appearance, where the oars dashed the water. Sometimes it was very bright, and afterwards as we rowed along, gradually lessened, till almost imperceptible, and then re-illumined. This we took notice of several times in the passage. When I got on board the vessel, I ordered a pail to be dipped up, full of sea-water, in which, on the water's being moved, a sparkling light appeared. I took a linnen cloth, and strained some of the water through it, and there was a like appearance on the cloth, which soon went off; but on rubbing the

cloth

cloth with my finger, it was renewed. I then carried the cloth to the light, but could not perceive any thing upon it which fhould caufe that appeaiance.

Several Gentlemen were of opinion, that the feparated particles of putrid, animal, and other bodies, floating on the furface of the fea, might caufe that appearance ; for putrid fifh, &c. they faid, will caufe it : And the fea-animals which have died, and other bodies putrified therein fince the creation, might afford a fufficient quantity of thefe particles to cover a confiderable portion of the furface of the fea ; which particles being differently difperfed, might account for the different degrees of light in the appearance above-mentioned. But this account feems liable to this obvious objection, That as putrid fifh, &c. make a luminous appearance without being moved or difturbed, it might be expected that the fuppofed putrid particles on the furface of the fea, fhould always appear luminous, where there is not a greater light ; and, confequently, that the whole furface of the fea, covered with thofe particles, fhould always, in dark nights, appear luminous, without being difturbed. But this is not fact.

Among the reft, I threw out my conjecture, That the faid appearance might be caufed by a great number of little animals, floating on the furface of the fea, which, on being difturbed, might, by expanding their finns, or otherwife moving themfelves, expofe fuch a part of their bodies as exhibits a luminous appearance, fomewhat in the manner of a glow-worm, or fire-fly : That thefe animals

may

may be more numerous in fome places than others ; and, therefore, that the appearance above-mentioned being fainter and ftronger in different places, might be owing to that : That certain circumftances of weather, &c. might invite them to the furface, on which, in a calm, they might fport themfelves and glow ; or in ftorms, being forced up, make the fame appearance.

There is no difficulty in conceiving that the fea may be ftocked with animalcula for this purpofe, as we find all Nature crowded with life. But it feems difficult to conceive that fuch fmall portions of matter, even if they were wholly luminous, fhould affect our fight ; much more fo, when it is fuppofed that only a part of them is luminous. But, if we confider fome other appearances, we may find the fame difficulty to conceive of them ; and yet we know they take place. For inftance, the flame of a candle, which, it is faid, may be feen four miles round. The light which fills this circle of eight miles diameter, was contained when it firft left the candle within a circle of half an inch diameter. If the denfity of light, in thefe circumftances, be as thofe circles to each other, that is, as the fquares of their diameters, the candle-light, when come to the eye, will be 1027709337600 times rarer than when it quitted the half inch circle. Now the aperture of the eye, through which the light paffes, does not exceed one-tenth of an inch diameter, and the portion of the leffer circle, which correfponds to this fmall portion of the greater circle, muft be proportionably, that

O o

is,

is, 1027709337600 times lefs than one-tenth of an inch; and yet this infinitely fmall point (if you will allow the expreffion) affords light enough to make it vifible four miles; or, rather, affords light fufficient to affect the fight at that diftance.

The fmallnefs of the animalcula is no objection then to this conjecture; for fuppofing them to be ten thoufand times lefs than the *minimum vifibile*, they may, notwithftanding, emit light enough to affect the eyes, and fo to caufe the luminous appearance aforefaid. This conjecture I fend you for want of fomething better.

Farther REMARKS *by a Gentleman of* New-York.

April 2, 1754.

Read at the Royal-
Society, *Dec.* 16,
1756.

ANY knowledge I have of the winds, and other changes which happen in the atmofphere, is fo very defective that it does not deferve the name; neither have I received any fatisfaction from the attempts of others on this fubject. It deferves then your thoughts, as a fubject in which you may diftinguifh yourfelf and be ufeful.

Your notion of fome things conducting heat or cold better than others, pleafes me, and I wifh you may purfue the fcent. If I remember right, Dr *Boerhaave*, in his

chy-

chymiftry, thinks that heat is propagated by the vibration
of a fubtle elaftic fluid, difperfed through the atmofphere
and through all bodies. Sir *Ifaac Newton* fays, there are
many phænomena to prove the exiftence of fuch a fluid;
and this opinion has my affent to it. I fhall only obferve
that it is effentially different from that which I call æther;
for æther, properly fpeaking, is neither a fluid nor elaftic;
its power confifts in re-acting any action communicated to
it, with the fame force it receives the action.

I long to fee your explication of water-fpouts, but I
muft tell you before hand, that it will not be eafy for
you to convince me that the principal phænomena were
not occafioned by a ftream of wind iffuing with great
force, my eyes and ears both concurring to give me this
fentiment, I could have no more evidence than to feel the
effects, which I had no inclination to do.

It furprifes me a little, that wind, generated by fermentation, is new to you, fince it may be every day obferved
in fermenting liquor. You know with what force fermenting liquors will burft the veffels which contain them,
if the generated wind have not vent; and with what
force it iffues on giving it a fmall vent, or by drawing the
cork of a bottle. Dr *Boerhaave* fays, that the fteam iffuing from fermenting liquors received through a very fmall
vent-hole, into the nofe, will kill as fuddenly and certainly as lightning. That air is generated by fermentation, I
think you will find fully proved in Dr *Hales's Analyfis of*

the

the air, in his Vegetable Statics. If you have not read the book, you have a new pleafure to come.

The folution you give to the objection I made from the contrary winds blowing from the oppofite fides of the mountains, from their being eddies, does not pleafe me, becaufe the extent of thefe winds is by far too large to be occafioned by an eddy. It is forty miles from *New-York* to our mountains, through which *Hudfon's River* paffes. The river runs twelve miles in the mountains, and from the North fide of the mountains it is about ninety miles to *Albany*. I have myfelf been on board a veffel more than once, when we have had a ftrong Northerly wind againft us, all the way from *New-York*, for two or three days. We have met veffels from *Albany*, who affured us, that, on the other fide of the mountains, they had, at the fame time, a ftrong continued Southerly wind againft them; and this frequently happens.

I have frequently feen both, on the river, in places where there could be no eddy-winds, and on the open fea, two veffels failing with contrary winds, within half a mile of each other; but this happens only in eafy winds, and generally calm in other places near thefe winds.

You have, no doubt, frequently obferved a fingle cloud pafs, from which a violent guft of wind iffues, but of no great extent. I have obferved fuch a guft make a lane through the woods, of fome miles in length, by laying the trees flat to the ground, and not above eight or ten chains in breadth. Though the violence of the wind be in the

fame

fame direction in which the cloud moves and precedes it, yet wind issues from all sides of it ; so that supposing the cloud move South-Easterly, those on the North-East side of it feel a South-West wind, and others on the South-West side, a North East. And where the cloud passes over, we frequently have a South-East wind from the hinder part of it, but none violent, except the wind in the direction in which the cloud moves. To shew what it is which prevents the wind from issuing out equally on all sides, is not an easy problem to me, and I shall not attempt to solve it ; but when you shall shew what it is which restrains the electrical fluid from spreading itself into the air surrounding it, when it rushes with great violence through the air along, or in the conductor, for a great extent in length, then I may hope to explain the other problem, and remove the difficulty we have in conceiving it.

Propofal of an EXPERIMENT *to meafure the time taken up by an Electric Spark, in moving through any given Space.* By J. A. *Efq; of* New-York.

Read at the Royal Society *Dec.* 26, 1756.

IF I remember right, the Royal Society made one experiment to difcover the velocity of the electric fire, by a wire of about four miles in length, fupported by filk, and by turning it forwards and backwards in a field, fo that the beginning and end of the wire were at only the diftance of two people, the one holding the *Leyden* bottle and the beginning of the wire, and the other holding the end of the wire and touching the ring of the bottle; but by this experiment no difcovery was made, except that the velocity was extremely quick.

As water is a conductor as well as metals, it is to be confidered whether the velocity of the electric fire might not be difcovered by means of water; whether a river, or lake, or fea, may not be made part of the circuit through which the electric fire paffes? inftead of the circuit all of wire, as in the above experiment.

Whether in a river, lake, or fea, the electric fire will not diffipate and not return to the bottle? or, will it proceed in ftrait lines through the water the fhorteft courfes poffible back to the bottle.

If the laft, then fuppofe one brook that falls into *Delaware* doth head very near to a brook that falls into *Schuyl-kill,*

kill, and let a wire be ftretched and fupported as before, from the head of the one brook to the head of the other, and let the one end communicate with the water, and let one perfon ftand in the other brook, holding the *Leyden* bottle, and let another perfon hold that end of the wire not in the water, and touch the ring of the bottle.—If the electric fire will go as in the laft queftion, then will it go down the one brook to *Delaware* or *Schuylkill,* and down one of them to their meeting, and up the other and the o-ther brook; the time of its doing this may poffibly be ob-fervable, and the further upwards the brooks are chofen, the more obfervable it would be.

Should this be not obfervable, then fuppofe the two brooks falling into *Safquehana* and *Delaware,* and proceed-ing as before, the electric fire may, by that means, make a circuit round the North Cape of *Virginia,* and go many hundreds of miles, and in doing that, it would feem it muft take fome obfervable time.

If ftill no obfervable time is found in that experiment, then fuppofe the brooks falling the one into the *Ohio,* and the other into *Safquehana,* or *Potomack,* in that the electric fire would have a circuit of fome thoufands of miles to go down *Ohio* to *Miffifippi,* to the Bay of *Mexico,* round *Flori-da,* and round the South Cape of *Virginia;* which, I think, would give fome obfervable time, and difcover exactly the velocity.

But if the electric fire diffipates, or weakens in the water, as I fear it does, thefe experiments will not anfwer.

An-

Anſwer to the foregoing ; *by* B. F.

Read at the Royal-Society, *Dec.* 23, 1756.

SUppoſe a tube of any length open at both ends, and containing a moveable wire of juſt the ſame length, that fills its bore. If I attempt to introduce the end of another wire into the ſame tube, it muſt be done by puſhing forward the wire it already contains ; and the inſtant I preſs and move one end of that wire, the other end is alſo moved ; and in introducing one inch of the ſame wire, I extrude, at the ſame time, an inch of the firſt, from the other end of the tube.

If the tube be filled with water, and I injeÄ an additional inch of water at one end, I force out an equal quantity at the other, in the very ſame inſtant.

And the water forced out at one end of the tube is not the very ſame water that was forced in at the other end at the ſame time, it was only in motion at the ſame time.

The long wire made uſe of in the experiment to diſcover the velocity of the eleÄric fluid, is itſelf filled with what we call its natural quantity of that fluid, before the hook of the *Leyden* bottle is applied to one end of it.

The outſide of the bottle being at the time of ſuch application, in contaÄ with the other end of the wire ; the

whole

whole quantity of electric fluid contained in the wire is, probably, put in motion at once.

For at the inftant the hook, connected with the infide of the bottle, *gives out*; the coating, or outfide of the bottle, *draws in* a portion of that fluid.

If fuch long wire contains precifely the quantity that the outfide of the bottle demands, the whole will move out of the wire to the outfide of the bottle, and the over quantity which the infide of the bottle contained, being exactly equal, will flow into the wire, and remain there, in the place of the quantity the wire had juft parted with to the outfide of the bottle.

But if the wire be fo long as that one-tenth (fuppofe) of its natural quantity is fufficient to fupply what the outfide of the bottle demands, in fuch cafe the outfide will only receive what is contained in one-tenth of the wire's length, from the end next to it; though the whole will move fo as to make room at the other end for an equal quantity iffuing, at the fame time, from the infide of the bottle.

So that this experiment only fhews the extream facility with which the electric fluid moves in metal; it can never determine the velocity.

And, therefore, the propofed experiment (though well imagined, and very ingenious) of fending the fpark round through a vaft length of fpace, by the waters of *Sufquehannah*, or *Potowmack*, and *Ohio*, would not afford the fatisfaction defired, though we could be fure that the motion of the electric fluid would be in that tract, and not under ground in the wet earth by the fhorteft way.

P p

An

An Account of the new-invented Pensylvanian FIRE-PLACES : *Wherein their Construction and Manner of Operation is particularly explained; their Advantages above every other Method of warming Rooms demonstrated ; and all Objections that have been raised against the Use of them, answered and obviated. With Directions for putting them up, and for using them to the best Advantage. And a Copper-Plate, in which the several Parts of the Machine are exactly laid down, from a Scale of equal Parts.*

By B. F. *first printed at* Philadelphia *in* 1745.

IN these Northern Colonies the inhabitants keep fires to sit by, generally seven months in the year ; that is, from the beginning of *October*, to the end of *April*; and, in some winters, near eight months, by taking in part of *September* and *May*.

Wood, our common fuel, which within these hundred years might be had at every man's door, must now be fetched near one hundred miles to some towns, and makes a very considerable article in the expence of families.

As therefore so much of the comfort and conveniency of our lives, for so great a part of the year, depends on the article of *fire* ; since fuel is become so expensive, and (as the country is more cleared and settled) will of course grow scarcer and dearer, any new proposal for saving the
wood,

wood, and for leffening the charge, and augmenting the benefit of fire, by fome particular method of making and managing it, may at leaft be thought worth confideration.

The new FIRE-PLACES are a late invention to that pur- pofe, of which this paper is intended to give a particular account.

That the reader may the better judge whether this me- thod of managing fire has any advantage over thofe here- tofore in ufe, it may be proper to confider both the old and new methods feparately and particularly, and after- wards make the comparifon.

In order to this, 'tis neceffary to underftand well, fome few of the properties of air and fire, *viz.*

1. Air is rarified by *heat*, and condens'd by *cold*; *i. e.* the fame quantity of air takes up more fpace when warm than when cold. This may be fhown by feveral very eafy experiments. Take any clear glafs bottle (a *Florence* flafk ftript of the ftraw is beft) place it before the fire, and as the air within is warmed and rarified, part of it will be driven out of the bottle; turn it up, place its mouth in a veffel of water, and remove it from the fire; then, as the air within cools and contraĉts, you will fee the water rife in the neck of the bottle, fupplying the place of juft fo much air as was driven out. Hold a large hot coal near the fide of the bottle, and as the air within feels the heat, it will again diftend and force out the water.——Or, fill a bladder half full of air, tie the neck tight, and lay it be- fore a fire as near as may be without fcorching the blad-

der;

der ; as the air within heats, you will perceive it to fwell and fill the bladder, till it becomes tight, as if full blown : Remove it to a cool place, and you will fee it fall gradually, till it becomes as lank as at firſt.

2. Air rarified and diſtended by heat, is * ſpecifically lighter than it was before, and will riſe in other air of greater denſity. As wood, oil, or any other matter ſpecifically lighter than water, if placed at the bottom of a veſſel of water, will riſe till it comes to the top ; ſo rarified air will riſe in common air, till it either comes to air of equal weight, or is by cold reduced to its former denſity.

A fire then being made in any chimney, the air over the fire is rarified by the heat, becomes lighter, and therefore immediately riſes in the funnel, and goes out ; the other air in the room (flowing towards the chimney) ſupplies its place, is rarified in its turn, and riſes likewiſe ; the place of the air thus carried out of the room, is ſupplied by freſh air coming in through doors and windows, or, if they be ſhut, through every crevice with violence, as may be ſeen by holding a candle to a key-hole : If the room be ſo tight as that all the crevices together will not ſupply ſo much air as is continually carried off, then, in a little time, the current up the funnel muſt flag, and the ſmoke being no longer driven up, muſt come into the room.

* Body or matter of any ſort, is ſaid to be *ſpecifically* heavier or lighter than other matter, when it has more or leſs ſubſtance or weight in the ſame dimenſions.

1. Fire,

1. Fire, (*i. e.* common fire) throws out light, heat, and fmoke (or fume.) The two firft move in right lines, and with great fwiftnefs, the latter is but juft feparated from the fuel, and then moves only as it is carried by the ftream of rarified air : And without a continual acceffion and receffion of air, to carry off the fmoaky fumes, they would remain crouded about the fire, and ftifle it.

2. Heat may be feparated from the fmoke as well as from the light, by means of a plate of iron, which will fuffer heat to pafs through it without the others.

3. Fire fends out its rays of heat, as well as rays of light, equally every way ; but the greateft fenfible heat is over the fire, where there is, befides the rays of heat fhot upwards, a continual rifing ftream of hot air, heated by the rays fhot round on every fide.

Thefe things being underftood, we proceed to confider the Fire-places heretofore in ufe, *viz.*

1. The large open fire-places ufed in the days of our fathers, and ftill generally in the country, and in kitchens.

2. The newer-fafhioned fire-places, with low breafts, and narrow hearths.

3. Fire-places with hollow backs, hearths and jams of iron, (defcribed by M. *Gauger*, in his tract entitled, *La Mechanique de Feu*) for warming the air as it comes into the room.

4. The *Holland* ftoves, with iron doors opening into the room.

5. The

5. The *German* ftoves, which have no opening in the room where they are ufed, but the fire is put in from fome other room, or from without.

6. Iron pots, with open charcoal fires, placed in the middle of a room.

1. The firft of thefe methods has generally the conveniency of two warm feats, one in each corner ; but they are fometimes too hot to abide in, and, at other times, incommoded with the fmoke ; there is likewife good room for the cook to move, to hang on pots, &c. Their inconveniences are, that they almoft always fmoke, if the door be not left open ; that they require a large funnel, and a large funnel carries off a great quantity of air, which occafions what is called a ftrong draft to the chimney, without which ftrong draft, the fmoke would come out of fome part or other of fo large an opening, fo that the door can feldom be fhut ; and the cold air fo nips the backs and heels of thofe that fit before the fire, that they have no comfort till either fcreens or fettles are provided (at a confiderable expence) to keep it off, which both cumber the room, and darken the fire fide. A moderate quantity of wood on the fire, in fo large a hearth, feems but little ; and, in fo ftrong and cold a draught, warms but little ; fo that people are continually laying on more. In fhort, 'tis next to impoffible to warm a room with fuch a fire place : And I fuppofe our anceftors never thought of warming rooms to fit in ; all they purpos'd was, to have a place to make a fire in, by which they might warm themfelves when cold.

2. Moft

2. Moſt of theſe old-faſhioned chimneys in towns and cities, have been, of late years, reduced to the ſecond ſort mentioned, by building jambs within them, narrowing the hearth, and making a low arch or breaſt. 'Tis ſtrange, methinks, that though chimneys have been ſo long in uſe, their conſtruction ſhould be ſo little underſtood till lately, that no workman pretended to make one which ſhould always carry off all the ſmoke, but a chimney-cloth was looked upon as eſſential to a chimney. This improvement, however, by ſmall openings and low breaſts, has been made in our days; and ſucceſs in the firſt experiments has brought it into general uſe in cities, ſo that almoſt all new chimnies are now made of that ſort, and much fewer bricks will make a ſtack of chimneys now than formerly. An improvement ſo lately made, may give us room to believe, that ſtill farther improvements may be found to remedy the inconveniencies yet remaining. For theſe new chimneys, though they keep rooms generally free from ſmoke, and, the opening being contracted, will allow the door to be ſhut, yet the funnel ſtill requiring a conſiderable quantity of air, it ruſhes in at every crevice ſo ſtrongly, as to make a continual whiſtling or howling, and it is very uncomfortable, as well as dangerous, to ſit againſt any ſuch crevice. Many colds are caught from this cauſe only, it being ſafer to ſit in the open ſtreet, for then the pores do all cloſe together, and the air does not ſtrike ſo ſharply againſt any particular part of the body.

The

The *Spaniards* have a proverbial saying,

If the wind blows on you through a hole,
Make your will, and take care of your soul.

Women, particularly, from this caufe, as they fit much in the houfe, get colds in the head, rheums and deflucti-ons, which fall into their jaws and gums, and have de-ftroyed early many a fine fet of teeth in thefe Northern co-lonies. Great and bright fires do alfo very much contri-bute to damage the eyes, dry and fhrivel the fkin, and bring on early the appearances of old age. In fhort, ma-ny of the difeafes proceeding from colds, as fevers, pleu-rifies, &c. fatal to very great numbers of people, may be afcribed to ftrong drawing chimneys, whereby, in fevere weather, a man is fcorched before, while he is froze be-hind *. In the mean time, very little is done by thefe

chimneys

---

* As the writer is neither phyfician nor philofopher, the reader may expect he fhould juftify thefe his opinions by the authority of fome that are fo. M. *Clare,* F.R.S. in his treatife of *The motion of fluids,* fays, *pag.* 246, &c. " And here it may be remarked, that it is more prejudicial to " health, to fit near a window or door, in a room where there are many " candles and a fire, than in a room without; for the confumption of air " thereby occafioned, will always be very confiderable, and this muft ne-" ceffarily be replaced by cold air from without. Down the chimney can " enter none, the ftream of warm air, always arifing therein, abfolutely " forbids it; the fupply muft therefore come in wherever other open-" ings fhall be found. If thefe happen to be fmall, *let thofe who fit near* " *them beware;* the fmaller the floodgate, the fmarter will be the ftream. " Was a man, even in a fweat, to leap into a cold bath, or jump from " his warm bed, in the intenfeft cold, even in a froft, provided he do not " continue over-long therein, and be in health when he does this, we fee " by experience that he gets no harm., If he fits a little while againft a " window, into which a fucceffive current of cold air comes, his pores are " clofed, and he gets a fever. In the firft cafe, the fhock the body en-dures

chimneys towards warming the room ; for the air round the fire-place, which is warmed by the direct rays from the fire, does not continue in the room, but is continually crouded and gathered into the chimney by the current of cold air coming behind it, and so is presently carried off.

" dures is general, uniform, and therefore less fierce ; in the other a single
" part, a neck or ear perchance, is attacked, and that with the greater
" violence probably, as it is done by a successive stream of cold air.   And
" the cannon of a battery, pointed against a single part of a bastion, will
" easier make a breach than were they directed to play singly upon the
" whole face, and will admit the enemy much sooner into the town."

That warm rooms, and keeping the body warm in winter, are means of preventing such diseases, take the opinion of that learned *Italian* physician *Antonio Porcio,* in the preface to his tract *de Militis Sanitate tuenda,* where, speaking of a particular wet and cold winter, remarkable at *Venice* for its sickliness, he says, *Popularis autem pleuritis quæ Venetiis sæviit mensibus* Dec. Jan. Feb. *ex cœli, aërisque inclementia facta est, quod non habeant hypocausta* [stove-rooms] *& quod non soliciti sunt* Itali *omnes de auribus, temporibus, collo, totoque corpore defendendis ab injuriis aëris; et tegmina domorum* Veneti *disponant parum inclinata, ut nives diutius permaneant super tegmina. E contra,* Germani, *qui experiuntur cœli inclementiam, perdidicere sese defendere ab aëris injuria. Tecta construunt multum inclinata, ut decidant nives.* Germani *abundant lignis, domusque* hypocaustis *; foris autem incedunt pannis, pellibus, gossipio, bene mehercule loricati atque muniti. In* Bavaria *interrogabam (curiositate motus videndi* Germaniam*) quot nam elapsis mensibus pleuritide vel peripneumonia fuissent absumti ; dicebant vix unus aut alter illis temporibus pleuritide fuit correptus.*

The great Dr *Boerhaave,* whose authority alone might be sufficient, in his Aphorisms mentions, as one antecedent cause of pleurisies, *a cold air, driven violently through some narrow passage upon the body, overheated by labour or fire.*

The Eastern physicians agree with the *Europeans* in this point ; witness the *Chinese* treatise entitled *Tchang seng,* i. e. *The art of procuring health and long life,* as translated in *Pere Du Halde*'s account of *China,* which has this passage. *As of all the passions which ruffle us, Anger does the most mischief, so of all the malignant affections of the air, a* wind that comes thro' any narrow passage, *which is cold and piercing,* is most dangerous ; *and coming upon us unawares, insinuates itself into the body, often causing grievous diseases. It should therefore be avoided, according to the advice of the ancient proverb, as carefully as the point of an arrow.* These mischiefs are avoided by the use of the new-invented fire-places, as will be shewn hereafter.

Q q   In

In both thefe forts of fire-places, the greateſt part of the heat from the fire is loſt; for as fire naturally darts heat every way, the back, the two jambs, and the hearth, drink up almoſt all that is given them, very little being reflected from bodies ſo dark, porous, and unpoliſhed ; and the upright heat, which is by far the greateſt, flies directly up the chimney. Thus five ſixths at leaſt of the heat (and confequently of the fewel) is waſted, and contributes nothing towards warming the room.

3. To remedy this, the Sieur *Gauger* gives, in his book entitled, *La Mechanique de Feu,* publiſhed in 1709, feven different conſtructions of the third fort of chimneys mentioned above, in which there are hollow cavities made by iron plates in the back, jambs, and hearth, through which plates the heat paſſing, warms the air in thoſe cavities, which is continually coming into the room freſh and warm. The invention was very ingenious, and had many conveniencies: The room was warmed in all parts, by the air flowing into it through the heated cavities : Cold air was prevented ruſhing through the crevices, the funnel being ſufficiently ſupplied by thoſe cavities : Much leſs fuel would ſerve, *&c.* But the firſt expence, which was very great ; the intricacy of the defign, and the difficulty of the execution, eſpecially in old chimnies, difcouraged the propagation of the invention ; ſo that there are, I ſuppoſe, very few ſuch chimnies now in uſe. [The upright heat, too, was almoſt all loſt in theſe, as in the common chimneys.]

4. The

4. The *Holland* iron ftove, which has a flue proceeding from the top, and a fmall iron door opening into the room, comes next to be confidered. Its conveniencies are, that it makes a room all over warm ; for the chimney being wholly clofed, except the flue of the ftove, very little air is required to fupply that, and therefore not much rufhes in at crevices, or at the door when it is opened. Little fewel ferves, the heat being almoft all faved ; for it rays out almoft equally from the four fides, the bottom and the top, into the room, and prefently warms the air around it, which being rarified, rifes to the cieling, and its place is fupplied by the lower air of the room, which flows gradually towards the ftove, and is there warmed, and rifes in its turn, fo that there is a continual circulation till all the air in the room is warmed. The air, too, is gradually changed, by the ftove-door's being in the room, through which part of it is continually paffing, and that makes thefe ftoves wholefomer, or at leaft pleafanter than the *German* ftoves, next to be fpoke of —But they have thefe inconveniences. There is no fight of the fire, which is in itfelf a pleafant thing. One cannot conveniently make any other ufe of the fire but that of warming the room. When the room is warm, people not feeing the fire, are apt to forget fupplying it with fuel till it is almoft out, then, growing cold, a great deal of wood is put in, which foon makes it too hot. The change of air is not carried on quite quick enough, fo that if any fmoke or ill fmell happens in the room, it is a long time before it is difcharged.

For

For thefe reafons the *Holland* ftove has not obtained much among the *Englifh* (who love the fight of the fire) unlefs in fome workfhops, where people are obliged to fit near windows for the light, and in fuch places they have been found of good ufe.

5. The *German* ftove is like a box, one fide wanting. It is compofed of five iron plates fcrued together, and fixed fo as that you may put the fuel into it from another room, or from the outfide of the houfe. It is a kind of oven re-verfed, its mouth being without, and body within the room that is to be warmed by it. This invention certainly warms a room very fpeedily and thoroughly with little fuel : No. quantity of cold air comes in at any crevice, becaufe there is no difcharge of air which it might fupply, there being no paffage into the ftove from the room. Thefe are its conveniencies.—Its inconveniencies are, That people have not even fo much fight or ufe of the fire as in the *Holland* ftoves, and are, moreover, obliged to breathe the fame un-chang'd air continually, mixed with the breath and perfpi-ration from one another's bodies, which is very difagree-able to thofe who have not been accuftomed to it.

6. Charcoal fires in pots, are ufed chiefly in the fhops of handicraftsmen. They warm a room (that is kept clofe, and has no chimney to carry off the warmed air) very fpeedily and uniformly ; but there being no draught to change the air, the fulphurous fumes from the coals [be they ever fo well kindled before they are brought in, there will be fome] mix with it, render it difagreeable, hurtful to

fome

fome conftitutions, and fometimes, when the door is long kept fhut, produce fatal confequences.

To avoid the feveral inconveniencies, and at the fame time retain all the advantages of other fire-places, was contrived the PENNSYLVANIA FIRE.PLACE, now to be defcribed.

This Machine confifts of

A bottom-plate, (i)   [*See the Plate annexed.*]

A back plate, (ii)

Two fide plates, (iii iii)

Two middle plates, (iv iv) which joined together, form a tight box, with winding paffages in it for warming the air.

A front plate, (v)

A top plate, (vi)

Thefe are all caft of iron, with mouldings or ledges where the plates come together, to hold them faft, and retain the mortar ufed for pointing to make tight joints. When the plates are all in their places, a pair of flender rods with fcrews, are fufficient to bind the whole very firmly together, as it appears in *Fig.* 2.

There are, moreover, two thin plates of wrought iron, *viz.* the fhutter, (vii) and the regifter, (viii) ; befides the fcrew-rods O P, all which we fhall explain in their order.

(i) The bottom plate or hearth-piece, is round before, with a rifing moulding that ferves as a fender to keep coals and afhes from coming to the floor, *&c.* It has two ears, F G, perforated to receive the fcrew-rods O P ; a long airhole, *a a,* through which the frefh outward air paffes up

into

into the air-box; and three fmoke-holes B C through
which the fmoke defcends and paffes away; all reprefented by dark fquares. It has alfo double ledges to receive
between them the bottom edges of the back plate, the
two fide-plates, and the two middle-plates. Thefe ledges
are about an inch afunder, and about half an inch high;
a profile of two of them joined to a fragment of plate, appears in *Fig.* 3.

(ii) The back plate is without holes, having only a pair
of ledges on each fide, to receive the back edges of the two

(iii iii) Side plates : Thefe have each a pair of ledges to
receive the fide-edges of the front plate, and a little fhoulder
for it to reft on; alfo two pair of ledges to receive the
fide edges of the two middle plates which form the air-
box; and an oblong air-hole near the top, through which
is difcharged into the room the air warmed in the air-box.
Each has alfo a wing or bracket, H and I, to keep in
falling brands, coals, &c, and a fmall hole, Q and R, for the
axis of the regifter to turn in.

(iv iv) The air-box is compofed of the two middle plates
D E and F G. The firft has five thin ledges or partitions
caft on it, two inches deep, the edges of which are received in fo many pair of ledges caft in the other. The tops
of all the cavities formed by thefe thin deep ledges, are
alfo covered by a ledge of the fame form and depth, caft
with them; fo that when the plates are put together, and
the joints luted, there is no communication between the

                                                             air-

air-box and the fmoke. In the winding paffages of this box, frefh air is warm'd as it paffes into the room.

(v) The front plate is arched on the under fide, and ornamented with foliages, &c. it has no ledges.

(vi) The top plate has a pair of ears, M N, anfwerable to thofe in the bottom plate, and perforated for the fame purpofe: It has alfo a pair of ledges running round the under fide, to receive the top edges of the front, back, and fide-plates. The air-box does not reach up to the top plate by two inches and half.

(vii) The fhutter is of thin wrought iron and light, of of fuch a length and breadth as to clofe well the opening of the fire-place. It is ufed to blow up the fire, and to fhut up and fecure it a nights. It has two brafs knobs for handles, *d d*, and commonly flides up and down in a groove, left, in putting up the fire-place, between the foremoft ledge of the fide-plates, and the face of the front plate ; but fome chufe to fet it afide when it is not in ufe, and apply it on occafion.

(viii) The regifter is alfo of thin wrought iron. It is placed between the back plate and air-box, and can, by means of the key S, be turned on its axis, fo as to lie in any pofition between level and upright.

The fcrew-rods O P are of wrought iron, about a third of an inch thick, with a button at bottom, and a fcrew and nut at top, and may be ornamented with two fmall braffes fcrewed on above the nuts.

To

To put this Machine to work,

1. A falfe back of four inch (or, in fhallow fmall chimneys, two inch) brick work is to be made in the chimney, four inches or more from the true back : From the top of this falfe back a clofing is to be made over to the breaft of the chimney, that no air may pafs into the chimney, but what goes under the falfe back, and up behind it.

2. Some bricks of the hearth are to be taken up, to form a hollow under the bottom plate ; acrofs which hollow runs a thin tight partition, to keep apart the air entering the hollow and the fmoke ; and is therefore placed between the air-hole and fmoke-holes.

3. A paffage is made, communicating with the outward air, to introduce that air into the fore part of the hollow under the bottom-plate, whence it may rife thro' the air-hole into the air-box.

4. A paffage is made from the back part of the hollow, communicating with the flue behind the falfe back : Through this paffage the fmoke is to pafs.

The fire-place is to be erected upon thefe hollows, by putting all the plates in their places, and fcrewing them together.

Its operation may be conceived by obferving the following

P R O-

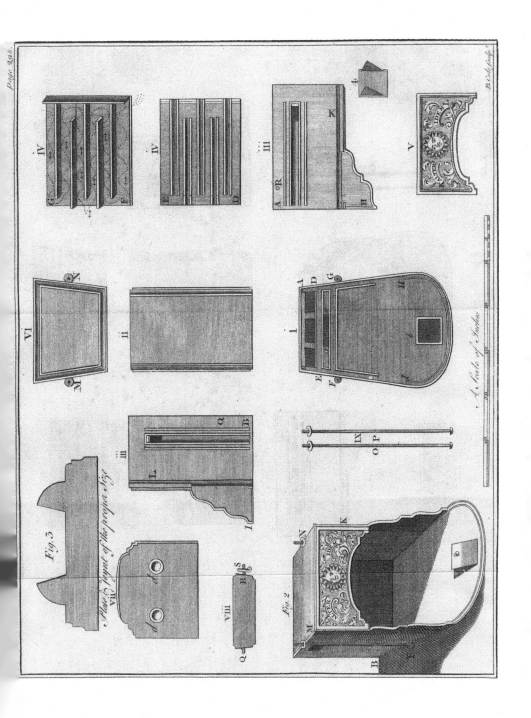

PROFILE *of the* CHIMNEY *and* FIRE-PLACE.

M The mantle-piece, or breast of the chimney.

C The funnel.

B The false back and closing.

E True back of the chimney.

T Top of the fire-place.

F The front of it.

A The place where the fire is made.

D The air-box.

K The hole in the side-plate, through which the warmed air is discharged out of the air-box into the room.

H The hollow filled with fresh air, entering at the passage *I*, and ascending into the air-box through the air-hole in the bottom plate near

G The partition in the hollow to keep the air and smoke apart.

P The passage under the false back and part of the hearth for the smoke.

The arrows show the course of the smoke.

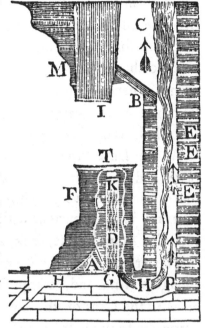

The fire being made at A, the flame and smoke will ascend and strike the top T, which will thereby receive a considerable heat. The smoke, finding no passage upwards, turns over the top of the air-box, and descends between it and the back plate to the holes at B, in the bottom plate, heating, as it passes, both plates of the air-box, and the said back plate; the front plate, bottom and side plates, are also all heated at the same time. The smoke proceeds in the passage that leads it under and behind the false back, and so rises into the chimney. The air of the room, warmed behind the back plate, and by the sides,

R r   front,

front, and top plates, becoming fpecifically lighter than the other air in the room, is obliged to rife ; but the clofure over the fire-place hindering it from going up the chimney, it is forced out into the room, rifes by the mantle-piece to the cieling, and fpreads all over the top of the room, whence being crouded down gradually by the ftream of newly-warm'd air that follows and rifes above it, the whole room becomes in a fhort time equally warmed.

At the fame time the air, warmed under the bottomplate, and in the air box, rifes and comes out of the holes in the fide-plates, very fwiftly if the door of the room be fhut, and joins its current with the ftream before mentioned, rifing from the fide, back, and top plates.

The air that enters the room through the air-box is frefh, though warm ; and, computing the fwiftnefs of its motion with the areas of the holes, it is found that near ten barrels of frefh air are hourly introduced by the air-box ; and by this means the air in the room is continually changed, and kept, at the fame time, fweet and warm.

It is to be obferved, that the entering air will not be warm at firft lighting the fire, but heats gradually as the fire encreafes.

A fquare opening for a trap-door fhould be left in the clofing of the chimney, for the fweeper to go up: The door may be made of flate or tin, and commonly kept clofe fhut, but fo placed as that turning up againft the back of the chimney when open, it clofes the vacancy behind the falfe back, and fhoots the foot, that falls in fweeping,

ing, out upon the hearth. This trap-door is a very convenient thing.

In rooms where much fmoking of tobacco is ufed, it is alfo convenient to have a fmall hole, about five or fix inches fquare, cut near the cieling through into the funnel: This hole muft have a fhutter, by which it may be clos'd or open'd at pleafure. When open, there will be a ftrong draught of air thro' it into the chimney, which will prefently carry off a cloud of fmoke, and keep the room clear: If the room be too hot likewife, it will carry off as much of the warm air as you pleafe, and then you may ftop it entirely, or in part, as you think fit. By this means it is, that the tobacco fmoke does not defcend among the heads of the company near the fire, as it muft do before it can get into common chimneys.

The Manner of ufing this FIRE-PLACE.

Your cord-wood muft be cut into three lengths; or elfe a fhort piece, fit for the fire-place, cut off, and the longer left for the kitchen or other fires. Dry hickery, or afh, or any woods that burn with a clear flame are rather to be chofen, becaufe fuch are lefs apt to foul the fmoke-paffages with foot; and flame communicates, with its light, as well as by contact, greater heat to the plates and room. But where more ordinary wood is ufed, half a dry faggot of brufh-wood, burnt at the firft making of fire in the morning is very advantageous, as it it immediately, by its fudden blaze, heats the plates, and warms the

R r 2 room

room (which with bad wood flowly kindling would not be done fo foon) and at the fame time, by the length of its flame, turning in the paffages, confumes and cleanfes away the foot that fuch bad fmoaky wood had produced therein the preceding day, and fo keeps them always free and clean.—When you have laid a little back log, and placed your billets on fmall dogs, as in common chimneys, and put fome fire to them, then flide down your fhutter as low as the dogs, and the opening being by that means contracted, the air rufhes in brifkly, and prefently blows up the flames. When the fire is fufficiently kindled, flide it up again *. In fome of thefe fire-places there is a little fix inch fquare trap-door of thin wrought iron or brafs, covering a hole of like dimenfions near the fore-part of the bottom-plate, which being by a ring lifted up towards the fire, about an inch, where it will be retained by two fpringing fides fixed to it perpendicularly, [*See the Plate, Fig.* 4,] the air rufhes in from the hollow under the bottom plate, and blows the fire. Where this is ufed, the fhutter ferves only to clofe the fire at nights. The more forward you can make your fire on the hearth-plate, not to be incommoded by the fmoke, the fooner and more will the room be warmed. At night when you go

* The fhutter is flid up and down in this manner, only in thofe fire-places which are fo made as that the diftance between the top of the arched opening, and the bottom plate, is the fame as the diftance between it and the top plate. Where the arch is higher, as it is in the draught annexed, (which is agreeable to the laft improvements) the fhutter is fet by, and applied occafionally; becaufe if it were made deep enough to clofe the whole opening when flid down, it would hide part of it when up.

to

to bed, cover the coals or brands with afhes as ufual; then take away the dogs, and flide down the fhutter clofe to the bottom-plate, fweeping a little afhes againft it, that no air may pafs under it; then turn the regifter, fo as very near to ftop the flue behind. If no fmoke then comes out at crevices into the room, it is right: If any fmoke is perceived to come out, move the regifter fo as to give a little draught, and it will go the right way.—Thus the room will be kept warm all night; for the chimney being almoft entirely ftopt, very little cold air, if any, will enter the room at any crevice. When you come to re-kindle the fire in the morning, turn open the regifter before you lift up the flider, otherwife, if there be any fmoke in the fire-place, it will come out into the room. By the fame ufe of the fhutter and regifter, a blazing fire may be prefently ftifled, as well as fecured, when you have occafion to leave it for any time; and at your return you will find the brands warm, and ready for a fpeedy re-kindling. The fhutter alone will not ftifle a fire, for it cannot well be made to fit fo exactly but that air will enter, and that in a violent ftream, fo as to blow up and keep alive the flames, and confume the wood, if the draught be not check'd by turning the regifter to fhut the flue behind. The regifter has alfo two other ufes. If you obferve the draught of air into your fire-place to be ftronger than is neceffary, (as in extream cold weather it often is) fo that the wood is confumed fafter than ufual; in that cafe, a quarter, half, or two thirds turn of the regifter, will check the violence of

the

the draught, and let your fire burn with the moderation you defire: And at the fame time both the fire-place and the room will be the warmer, becaufe lefs cold air will enter and pafs through them.—And if the chimney fhould happen to take fire, (which indeed there is very little danger of, if the preceding direction be obferved in making fires, and it be well fwept once a year; for, much lefs wood being burnt, lefs foot is proportionably made; and the fuel being foon blown into flame by the fhutter (or the trap-door bellows) there is confequently lefs fmoke from the fuel to make foot; then, though the funnel fhould be foul, yet the fparks have fuch a crooked up and down round about way to go, that they are out before they get at it. I fay, if ever it fhould be on fire, a turn of the regifter fhuts all clofe, and prevents any air going into the chimney, and fo the fire may eafily be ftifled and maftered.

The *Advantages of this* FIRE-PLACE.

Its advantages above the common fire-places are,

1. That your whole room is equally warmed, fo that people need not croud fo clofe round the fire, but may fit near the window, and have the benefit of the light for reading, writing, needle-work, &c. They may fit with comfort in any part of the room, which is a very confiderable advantage in a large family, where there muft often be two fires kept, becaufe all cannot conveniently come at one.

2. If you fit near the fire, you have not that cold draught of uncomfortable air nipping your back and heels, as when

before

before common fires, by which many catch cold, being fcorched before, and, as it were, froze behind.

3 If you fit againft a crevice, there is not that fharp draught of cold air playing on you as in rooms where there are fires in the common way; by which many catch cold, whence proceed coughs *, catarrhs, tooth-achs, fevers, pleurifies, and many other difeafes.

4. In cafe of ficknefs they make moft excellent nurfing rooms; as they conftantly fupply a fufficiency of frefh air, fo warmed at the fame time as to be no way inconvenient or dangerous. A fmall one does well in a chamber; and, the chimneys being fitted for it, it may be removed from one room to another, as occafion requires, and fixed in half an hour. The equal temper, too, and warmth, of the air of the room, is thought to be particularly advantageous in fome diftempers; for it was obferved in the winters of 1730 and 1736, when the fmall-pox fpread in *Pennfylvania*, that very few children of the *Germans* died of that diftemper in proportion to thofe of the *Englifh*; which was afcribed, by fome, to the warmth and equal temper of air in their ftove-rooms, which made the difeafe as favourable as it commonly is in the *Weft-Indies*. But this conjecture we fubmit to the judgment of phyficians.

---

* My Lord *Molefworth*, in his account of *Denmark*, fays, " That " few or none of the people there, are troubled with coughs, catarrhs, " confumptions, or fuch like difeafes of the lungs; fo that in the midft " of winter in the churches, which are very much frequented, there is " no noife to interrupt the attention due to the preacher. I am perfuad- " ed (fays he) their *warm ftoves* contribute to their freedom from thefe " kind of maladies." pag. 91.

5. In

5. In common chimneys, the ſtrongeſt heat from the fire, which is upwards, goes directly up the chimney, and is loſt ; and there is ſuch a ſtrong draught into the chimney, that not only the upright heat, but alſo the back, ſides, and downward heats, are carried up the chimney by that draught of air; and the warmth given before the fire by the rays that ſtrike out towards the room, is continually driven back, crouded into the chimney, and carried up by the ſame draught of air. But here the upright heat ſtrikes and heats the top plate, which warms the air above it, and that comes into the room. The heat likewiſe, which the fire communicates to the ſides, back, bottom, and air-box, is all brought into the room; for you will find a conſtant current of warm air coming out of the chimney corner into the room. Hold a candle juſt under the mantle-piece, or breaſt of your chimney, and you will ſee the flame bent outwards: By laying a piece of ſmoaking paper on the hearth, on either ſide, you may ſee how the current of air moves, and where it tends, for it will turn and carry the ſmoke with it.

6. Thus as very little of the heat is loſt, when this fire-place is uſed, *much leſs wood** will ſerve you, which is a conſiderable advantage where wood is dear.

7. When

* People who have uſed theſe fire-places, differ much in their accounts of the wood ſaved by them. Some ſay five-ſixths, others three-fourths, and others much leſs. This is owing to the great difference there was in their former fires; ſome (according to the different circumſtances of their rooms and chimnies) having been uſed to make very large, others
midd-

7. When you burn candles near this fire-place, you will find that the flame burns quite upright, and does not blare and run the tallow down, by drawing towards the chimney, as againſt common fires.

8. This fire-place cures moſt ſmoaky chimneys, and thereby preſerves both the eyes and furniture.

9. It prevents the fouling of chimneys; much of the lint and duſt that contributes to foul a chimney being, by the low arch, obliged to paſs through the flame, where it is conſumed. Then, leſs wood being burnt, there is leſs ſmoke made. Again, the ſhutter, or trap-bellows, ſoon blowing the wood into a flame, the ſame wood does not yield ſo much ſmoke as if burnt in a common chimney : For as ſoon as flame begins, ſmoke, in proportion, ceaſes.

10. And if a chimney ſhould be foul, it is much leſs likely to take fire. If it ſhould take fire, it is eaſily ſtifled and extinguiſhed.

11. A fire may be very ſpeedily made in this fire-place, by the help of the ſhutter, or trap-bellows, as aforeſaid.

12. A fire may be ſoon extinguiſhed, by cloſing it with the ſhutter before, and turning the regiſter behind, which will ſtifle it, and the brands will remain ready to rekindle.

13. The room being once warm, the warmth may be retained in it all night.

middling, and others, of a more ſparing temper, very ſmall ones: While in theſe fire places (their ſize and draught being nearly the ſame) the conſumption is more equal. I ſuppoſe, taking a number of families together, that two thirds, or half the wood, at leaſt, is ſaved. My common room, I know, is made twice as warm as it uſed to be, with a quarter of the wood I formerly conſumed there.

And

14. And, laftly, the fire is fo fecured at night, that not one fpark can fly out into the room to do damage.

With all thefe conveniencies, you do not lofe the pleafant fight nor ufe of the fire, as in the *Dutch* ftoves, but may boil the tea-kettle, warm the flat-irons, heat heaters, keep warm a difh of victuals, by fetting it on the top, &c.

Objections *anfwered.*

There are fome objections commonly made by people that are unacquainted with thefe fire-places, which it may not be amifs to endeavour to remove, as they arife from prejudices which might otherwife obftruct, in fome degree, the general ufe of this beneficial machine. We frequently hear it faid, *They are of the nature of* Dutch *ftoves; ftoves have an unpleafant fmell; ftoves are unwholefome;* and, *warm rooms make people tender, and apt to catch cold.*— As to the firft, that they are of the nature of *Dutch* ftoves, the defcription of thofe ftoves, in the beginning of this paper, compared with that of thefe machines, fhows that there is a moft material difference, and that thefe have vaftly the advantage, if it were only in the fingle article of the admiffion and circulation of the frefh air. But it muft be allowed there may have been fome caufe to complain of the offenfive fmell of iron ftoves. This fmell, however, never proceeded from the iron itfelf, which, in its nature, whether hot or cold, is one of the fweeteft of metals, but from the general uncleanly manner of ufing thofe ftoves. If they are kept clean, they are as fweet as an

iron

ironing-box, which, though ever so hot, never offends the smell of the nicest Lady : But it is common to let them be greased, by setting candlesticks on them, or otherwise ; to rub greasy hands on them ; and, above all, to spit upon them, to try how hot they are, which is an inconsiderate filthy unmannerly custom ; for the slimy matter of spittle drying on, burns and fumes when the stove is hot, as well as the grease, and smells most nauseously, which makes such close stove-rooms, where there is no draught to carry off those filthy vapours, almost intolerable to those that are not from their infancy accustomed to them. At the same time nothing is more easy than to keep them clean ; for when by any accident they happen to be fouled, a lee made of ashes and water, with a brush, will scour them perfectly ; as will also a little strong soft soap and water.

That hot iron of itself gives no offensive smell, those know very well who have (as the writer of this has) been present at a furnace when the workmen were pouring out the flowing metal to cast large plates, and not the least smell of it to be perceived. That hot iron does not, like lead, brass, and some other metals, give out unwholesome vapours, is plain from the general health and strength of those who constantly work in iron, as furnace-men, forge-men, and smiths ; that it is in its nature a metal perfectly wholesome to the body of man, is known from the beneficial use of chalybeate or iron mine-waters ; from the good done by taking steel filings in several disorders ; and that even the smithy water in which hot irons are quench-

ed,

ed, is found advantageous to the human conftitution.—
The ingenious and learned Dr *Defagutiers*, to whofe in-
ftructive writings the contriver of this machine acknow-
ledges himfelf much indebted, relates an experiment he
made, to try whether heated iron would yield unwholefome vapours: He took a cube of iron, and having given
it a very great heat, he fixed it fo to a receiver, exhaufted
by the air-pump, that all the air rufhing in to fill the receiver, fhould firft pafs through a hole in the hot iron. He
then put a fmall bird into the receiver, who breathed that
air without any inconvenience, or fuffering the leaft difor-
der. But the fame experiment being made with a cube of
hot brafs, a bird put into that air died in a few minutes.
Brafs, indeed, ftinks even when cold, and much more
when hot; lead, too, when hot, yields a very unwholefome fteam; but iron is always fweet, and every way taken is wholefome and friendly to the human body—except
in weapons.

*That warm rooms make people tender, and apt to catch
cold*, is a miftake as great as it is (among the *Englifh*) general. We have feen in the preceding pages how the
common rooms are apt to give colds; but the writer of
this paper may affirm from his own experience, and that
of his family and friends who have ufed warm rooms for
thefe four winters paft, that by the ufe of fuch rooms,
people are rendered *lefs liable* to take cold, and, indeed,
actually hardened. If fitting warm in a room made one
fubject to take cold on going out, lying warm in bed fhould,

by

by a parity of reason, produce the same effect when we rise. Yet we find we can leap out of the warmest bed naked, in the coldest morning, without any such danger ; and in the same manner out of warm cloaths into a cold bed. The reason is, that in these cases the pores all close at once, the cold is shut out, and the heat within augmented, as we soon after feel by the glowing of the flesh and skin. Thus no one was ever known to catch cold by the use of the cold bath : And are not cold baths allowed to harden the bodies of those that use them ? Are they not therefore frequently prescribed to the tenderest constitutions ? Now every time you go out of a warm room into the cold freezing air, you do as it were plunge into a cold bath, and the effect is in proportion the same ; for (though perhaps you may feel somewhat chilly at first) you find in a little time your bodies hardened and strengthened, your blood is driven round with a brisker circulation, and a comfortable steady uniform inward warmth succeeds that equal outward warmth you first received in the room. Farther to confirm this assertion, we instance the *Swedes*, the *Danes*, and the *Russians* : These nations are said to live in rooms, compared to ours, as hot as ovens * ; yet where are the hardy

---

* Mr *Boyle*, in his experiments and observations upon cold, *Shaw's Abridgement, Vol.* I. p. 684, says, " 'Tis remarkable, that while the cold " has strange and tragical effects at *Moscow*, and elsewhere, the *Russians* " and *Livonians* should be exempt from them, who accustom themselves " to pass immediately from a great degree of heat, to as great an one of " cold, without receiving any visible prejudice thereby. I remember be- " ing told by a person of unquestionable credit, that it was a common " practice

hardy foldiers, though bred in their boafted cool houfes,
that can, like thefe people, bear the fatigues of a winter
campaign in fo fevere a climate, march whole days to the
neck in fnow, and at night entrench in ice, as they do?

The mentioning of thofe Northern nations, puts me in
mind of a confiderable *public advantage* that may arife
from the general ufe of thefe fire-places.   It is obfervable,
that though thofe countries have been well inhabited for
many ages, wood is ftill their fuel, and yet at no very great
price ; which could not have been, if they had not univerfally ufed ftoves, but confumed it as we do, in great quantities, by open fires.   By the help of this faving invention
our wood may grow as faft as we confume it, and our pofterity may warm themfelves at a moderate rate, without
being obliged to fetch their fuel over the *Atlantick*; as, if
pit-coal fhould not be here difcovered, (which is an uncertainty) they muft neceffarily do.

We leave it to the *political arithmetician* to compute
how much money will be faved to a country, by its fpend-
ing two-thirds lefs of fuel; how much labour faved in
cutting and carriage of it ; how much more land may be
cleared by cultivation ; how great the profit by the additional quantity of work done, in thofe trades particularly that

" practice among them, to go from a hot ftove, into cold water ; the fame
" was alfo affirmed to me by another who refided at *Mofcow.*   This tra-
" dition is likewife abundantly confirmed by *Olearius*   "'Tis a furprizing
" thing, fays he, *to fee how far the* Ruffians *can endure heat ; and how,*
" *when it makes them ready to faint, they can go out of their ftoves, ftark*
" *naked, both men and women, and throw themfelves into cold water ; and*
" *even in winter wallow in the fnow.*"

do

do not exercife the body fo much, but that the workfolks are obliged to run frequently to the fire to warm themfelves: And to phyficians to fay, how much healthier thick-built towns and cities will be, now half fuffocated with fulphury fmoke, when fo much lefs of that fmoke fhall be made, and the air breathed by the inhabitants be confequently fo much purer. Thefe things it will fuffice juft to have mentioned ; let us proceed to give fome neceffary directions to the workman who is to fix or fet up thefe fire-places.

Directions *to the* Bricklayer.

The chimney being firft well fwept and cleanfed from foot,. &c. lay the bottom plate down on the hearth, in the place where the fire-place is to ftand, which may be as forward as the hearth will allow Chalk a line from one of its back corners round the plate to the other corner, that you may afterwards know its place when you come to fix it ; and from thofe corners, two parallel lines to the back of the chimney : Make marks alfo on each fide, that you may know where the partition is to ftand, which is to prevent any communication between the air and fmoke. Then removing the plate, make a hollow under it and beyond it, by taking up as many of the bricks or tiles as you can, within your chalked lines, quite to the chimney back. Dig out fix or eight inches deep of the earth or rubbifh, all the breadth and length of your hollow ; then make a paffage of four inches fquare (if the place will allow fo

much)

much) leading from the hollow to some place communicating with the outer air; by *outer air* we mean air without the room you intend to warm. This passage may be made to enter your hollow on either side, or in the fore part, just as you find most convenient, the circumstances of your chimney considered. If the fire-place is to be put up in a chamber, you may have this communication of outer air from the stair-case; or sometimes more easily from between the chamber floor and the cieling of the lower room, making only a small hole in the wall of the house entering the space betwixt those two joists with which your air-passage in the hearth communicates. If this air-passage be so situated as that mice may enter it, and nestle in the hollow, a little grate of wire will keep them out. This passage being made, and, if it runs under any part of the earth, tiled over securely, you may proceed to raise your false back. This may be of four inches or two inches thickness, as you have room, but let it stand at least four inches from the true chimney-back. In narrow chimnies this false back runs from jamb to jamb, but in large old fashioned chimnies, you need not make it wider than the back of the fire-place. To begin it, you may form an arch nearly flat, of three bricks end to end, over the hollow, to leave a passage the breadth of the iron fire-place, and five or six inches deep, rounding at bottom, for the smoke to turn and pass under the false back, and so behind it up the chimney. The false back is to rise till it is as high as the breast of the chimney, and then to close over

to

to the breaft * ; always obferving, if there is a wooden mantle-tree, to clofe above it. If there is no wood in the breaft, you may arch over and clofe even with the lower part of the breaft By this clofing the chimney is made tight, that no air or fmoke can pafs up it, without going under ihe falfe back. Then from fide to fide of your hollow, againft the marks you made with chalk, raife a tight partition, brick-on-edge, to feparate the air from the fmoke, bevelling away to half an inch the brick that comes juft under the air-hole, that the air may have a free paffage up into the air-box : Laftly, clofe the hearth over that part of the hollow that is between the falfe back and the place of the bottom plate, coming about half an inch under the plate, which piece of hollow hearth may be fupported by a bit or two of old iron hoop ; then is your chimney fitted to receive the fire-place.

To fet it, lay firft a little bed of mortar all round the edges of the hollow, and over the top of the partition : Then lay down your bottom plate in its place (with the rods in it) and tread it till it lies firm. Then put a little fine mortar (made of loam and lime with a little hair) into its joints, and fet in your back plate, leaning it for the prefent againft the falfe back : Then fet in your air-box, with a little mortar in its joints : Then put in the two fides clofing them up againft the air-box with mortar in their grooves, and fixing at the fame time your regifter : Then bring up

---

* See pag. 302, where the trap-door is defcribed that ought to be in this clofing.

your

your back to its place, with mortar in its grooves, and that will bind the fides together. Then put in your front plate, placing it as far back in the groove as you can, to leave room for the fliding plate : Then lay on your top plate, with mortar in its grooves alfo, fcrewing the whole firmly together by means of the rods. The capital letters A B D E, &c. in the annexed cut, fhew the correfponding parts of the feveral plates. Laftly, the joints being pointed all round on the outfide, the fire-place is fit for ufe.

When you make your firft fire in it, perhaps if the chimney be thoroughly cold, it may not draw, the work too being all cold and damp. In fuch cafe, put firft a few fhovels of hot coals in the fire-place, then lift up the chimney-fweeper's trap-door, and putting in a fheet or or two of flaming paper, fhut it again, which will fet the chimney a drawing immediately, and when once it is filled with a column of warm air, it will draw ftrongly and con- tinually.

The drying of the mortar and work by the firft fire, may fmell unpleafantly, but that will foon be over.

In fome fhallow chimneys, to make more room for the falfe back and its flue, four inches or more of the chimney back may be picked away.

Let the room be made as tight as conveniently it may be, fo will the outer air that muft come in to fupply the room and draught of the fire, be all obliged to enter thro' the paffage under the bottom-plate, and up through the air box, by which means it will not come cold to your

backs

backs, but be warmed as it comes in, and mixed with the warm air round the fire-place before it spreads in the room.

But as a great quantity of cold air, in extreme cold weather especially, will presently enter a room if the door be carelesly left open, it is good to have some contrivance to shut it, either by means of screw hinges, a spring, or a pulley.

When the pointing in the joints is all dry and hard, get some powder of black-lead, (broken bits of black-lead crucibles from the silver-smiths, pounded fine, will do) and mixing it with a little rum and water, lay it on, when the plates are warm, with a hard brush, over the top and front-plates, part of the side and bottom-plates, and over all the pointing; and, as it dries, rub it to a gloss with the same brush, so the joints will not be discerned, but it will look all of a piece, and shine like new iron. And the false back being plaister'd and white-wash'd, and the hearth redden'd, the whole will make a pretty appearance. Before the black-lead is laid on, it would not be amiss to wash the plates with strong lee and a brush, or soap and water, to cleanse them from any spots of grease or filth that may be on them. If any grease should afterwards come on them, a little wet ashes will get it out.

If it be well set up, and in a tolerable good chimney, smoke will draw in from as far as the fore part of the bottom plate, as you may try by a bit of burning paper.

People

People are at firſt apt to make their rooms too warm, not imagining how little a fire will be ſufficient. When the plates are no hotter than that one may juſt bear the hand on them, the room will generally be as warm as you deſire it.

Soon after the foregoing piece was publiſhed, ſome per-ſons in England, *in imitation of Mr* Franklin's *invention, made what they call* Penſylvania Fire-places, *with improve-ments ; the principal of which pretended improvements is a contraction of the paſſages in the air-box originally de-ſigned for admitting a quantity of freſh air, and warming it as it entered the room. The contracting theſe paſſages, gains indeed more room for the grate, but in a great mea-ſure defeats their intention. For if the paſſages in the air-box do not greatly exceed in dimenſions the amount of all the crevices by which cold air can enter the room, they will not conſiderably prevent, as they were intended to do, the entry of cold air through theſe crevices.*

LET-

LETTER XXIV.

FROM

BENJAMIN FRANKLIN, *Esq* of *Philadelphia*,

TO

Dr *L.*——, at *Charles-Town*, *South-Carolina*.

S I R, *Philadelphia, March* 18, 1755.

I SEND you enclofed a paper containing fome new experiments I have made, in purfuance of thofe by Mr *Canton* that are printed with my laft letters. I hope thefe, with my explanation of them, will afford you fome entertainment *.

In anfwer to your feveral enquiries. The tubes and globes we ufe here, are chiefly made here. The glafs has a greenifh caft, but is clear and hard, and, I think, better for eleftrical experiments than the white glafs of *London*, which is not fo hard. There are certainly great differences in glafs. A white globe I had made here fome years fince, would never, by any means, be excited. Two of my friends tried it, as well as myfelf, without fuccefs. At length

---

* See page 155, for the paper here mentioned.

length, putting it on an electric stand, a chain from the prime-conductor being in contact with it, I found it had the properties of a non-electric; for I could draw sparks from any part of it, though it was very clean and dry.

All I know of *Domien*, is, that by his own account he was a native of *Transylvania*, of *Tartar* descent, but a Priest of the *Greek* church: He spoke and wrote *Latin* very readily and correctly. He set out from his own country with an intention of going round the world, as much as possible by land. He travelled through *Germany*, *France*, and *Holland*, to *England*. Resided some time at *Oxford*. From *England* he came to *Maryland*; thence went to *New-England*; returned by land to *Philadelphia*; and from hence travelled through *Maryland*, *Virginia*, and *North-Carolina* to you. He thought it might be of service to him in his travels to know something of Electricity. I taught him the use of the tube; how to charge the *Leyden* phial, and some other experiments. He wrote to me from *Charles-Town*, that he had lived eight hundred miles upon Electricity, it had been meat, drink, and cloathing to him. His last letter to me was, I think, from *Jamaica*, desiring me to send the tubes you mention, to meet him at the *Havanah*, from whence he expected to get a passage to *La Vera Cruz*; designed travelling over land through *Mexico* to *Acapulco*; thence to get a passage to *Manilla*, and so through *China*, *India*, *Persia*, and *Turkey* home to his own country; proposing to support himself chiefly by Electricity. A strange project! But he was, as you ob-

serve,

ferve, a very fingular character. I was forry the tubes did not get to the *Havanah* in time for him: If they are ftill in being, pleafe to fend for them, and accept of them. What became of him afterwards I have never heard. He promifed to write to me as often as he could on his journey, and as foon as he fhould get home after finifhing his tour. It is now feven years fince he was here. If he is ftill in *New Spain*, as you imagine from that loofe report, I fuppofe it muft be that they confine him there, and prevent his writing : but I think it more likely that he may be dead.

The queftions you afk about the pores of glafs, I cannot anfwer otherwife, than that I know nothing of their nature ; and fuppofitions, however ingenious, are often mere miftakes. My hypothefis, that they were fmaller near the middle of the glafs, too fmall to admit the paffage of Electricity, which could pafs through the furface till it came near the middle, was certainly wrong : For foon after I had written that letter, I did, in order to *confirm* the hypothefis, (which indeed I ought to have done before I wrote it) make an experiment. I ground away five-fixths of the thicknefs of the glafs, from the fide of one of my phials, expecting that the fuppofed denfer part being fo removed, the electric fluid might come through the remainder of the glafs, which I had imagined more open ; but I found myfelf miftaken. The bottle charged as well after the grinding as before. I am now, as much as ever, at a lofs to know how or where the quantity of electric fluid, on the pofitive fide of the glafs, is difpofed of.

As

As to the difference of conductors, there is not only this, that fome will conduct Electricity in fmall quantities, and yet do not conduct it faft enough to produce the fhock ; but even among thofe that will conduct a fhock, there are fome that do it better than others. Mr *Kinnerfley* has found, by a very good experiment, that when the charge of a bottle hath an opportunity of paffing two ways, *i. e.* ftrait through a trough of water ten feet long, and fix inches fquare; or round about through twenty feet of wire, it paffes through the wire, and not through the water, though that is the fhorteft courfe; the wire being the better conductor. When the wire is taken away, it paffes through the water, as may be felt by a hand plunged in the water ; but it cannot be felt in the water when the wire is ufed at the fame time. Thus, though a fmall vial containing water will give a fmart fhock, one containing the fame quantity of mercury will give one much ftronger, the mercury being the better conductor ; while one containing oil only, will fcarce give any fhock at all.

Your queftion, how I came firft to think of propofing the experiment of drawing down the lightning, in order to afcertain its famenefs with the electric fluid, I cannot anfwer better than by giving you an extract from the minutes I ufed to keep of the experiments I made, with memorandums of fuch as I purpofed to make, the reafons for making them, and the obfervations that arofe upon them, from which minutes my letters were afterwards drawn.

By

By this extract you will fee that the thought was not fo much " an out-of-the-way one," but that it might have occurred to any electrician.

" *Nov.* 7, 1749. Electrical fluid agrees with lightning
" in thefe particulars : 1. Giving light. 2. Colour of the
" light. 3. Crooked direction. 4. Swift motion.
" 5. Being conducted by metals. 6. Crack or noife in
" exploding. 7. Subfifting in water or ice. 8. Rend-
" ing bodies it paffes through. 9. Deftroying animals.
" 10. Melting metals. 11. Firing inflammable fub-
" ftances. 12. Sulphureous fmell.—The electric fluid is
" attracted by points.—We do not know whether this
" property is in lightning.—But fince they agree in all
" the particulars wherein we can already compare them,
" is it not probable they agree likewife in this ?—Let the
" experiment be made."

I wifh I could give you any fatisfaction in the article of clouds.   I am ftill at a lofs about the manner in which they become charged with Electricity ; no hypothefis I have yet formed perfectly fatisfying me.   Some time fince, I heated very hot a brafs plate, two feet fquare, and placed it on an electric ftand.   From the plate a wire extended horizontally four or five feet, and, at the end of it, hung, by linnen threads, a pair of cork balls.   I then repeatedly fprinkled water over the plate, that it might be raifed from it in vapour, hoping that if the vapour either carried off the electricity of the plate, or left behind it that of the water, (one of which I fuppofed it muft do, if, like the clouds,

U u

it

it became electrifed itfelf, either pofitively or negatively)
I fhould perceive and determine it by the feparation of the
balls, and by finding whether they were pofitive or negative; but no alteration was made at all, nor could I
perceive that the fteam was itfelf electrifed, though I have
ftill fome fufpicion that the fteam was not fully examined,
and I think the experiment fhould be repeated.   Whether
the firft ftate of electrifed clouds is pofitive or negative, if I
could find the caufe of that, I fhould be at no lofs about the
other, for either is eafily deduced from the other, as one
ftate is eafily produced by the other.   A ftrongly pofitive
cloud may drive out of a neighbouring cloud much of its
natural quantity of the electric fluid, and, paffing by it,
leave it in a negative ftate.   In the fame way, a ftrongly
negative cloud may occafion a neighbouring cloud to draw
into itfelf from others, an additional quantity, and, paffing
by it, leave it in a pofitive ftate.   How thefe effects may
be produced, you will eafily conceive, on perufing and
confidering the experiments in the enclofed paper : And
from them too it appears probable, that every change from
pofitive to negative, and from negative to pofitive, that, during a thunder guft, we fee in the cork-balls annexed to
the apparatus, is not owing to the prefence of clouds in the
fame ftate, but often to the abfence of pofitive or negative
clouds, that, having juft paffed, leave the rod in the oppofite ftate.

The knocking down of the fix men was performed with
two of my large jarrs not fully charged.   I laid one end of
my

my difcharging rod upon the head of the firft; he laid his hand on the head of the fecond; the fecond his hand on the head of the third, and fo to the laft, who held, in his hand, the chain that was connected with the outfide of the jarrs. When they were thus placed, I applied the other end of my rod to the prime-conductor, and they all dropt together. When they got up, they all declared they had not felt any ftroke, and wondered how they came to fall; nor did any of them either hear the crack, or fee the light of it. You fuppofe it a dangerous experiment; but I had once fuffered the fame myfelf, receiving, by accident, an equal ftroke through my head, that ftruck me down, without hurting me: And I had feen a young woman that was about to be electrified through the feet, (for fome indifpofition) receive a greater charge through the head, by inadvertently ftooping forward to look at the placing of her feet, till her forehead (as fhe was very tall) came too near my prime-conductor: She dropt, but inftantly got up again, complaining of nothing. A perfon fo ftruck, finks down doubled, or folded together as it were, the joints lofing their ftrength and ftiffnefs at once, fo that he drops on the fpot where he ftood, inftantly, and there is no previous ftaggering, nor does he ever fall lengthwife. Too great a charge might, indeed, kill a man, but I have not yet feen any hurt done by it. It would certainly, as you obferve, be the eafieft of all deaths.

The experiment you have heard fo imperfect an account of, is merely this.—I electrified a filver pint cann,

on

on an electric stand, and then lowered into it a cork ball, of about an inch diameter, hanging by a silk string, till the cork touched the bottom of the cann. The cork was not attracted to the inside of the cann as it would have been to the outside, and though it touched the bottom, yet, when drawn out, it was not found to be electrified by that touch, as it would have been by touching the outside. The fact is singular. You require the reason ; I do not know it. Perhaps you may discover it, and then you will be so good as to communicate it to me *. I find a frank acknowledgment of one's ignorance is not only the easiest way to get rid of a difficulty, but the likeliest way to obtain information, and therefore I practice it : I think it an honest policy. Those who affect to be thought to know every thing, and so undertake to explain every thing, often remain long ignorant of many things that others could and would instruct them in, if they appeared less conceited.

The treatment your friend has met with is so common, that no man who knows what the world is, and ever has been, should expect to escape it. There are every where a number of people, who, being totally destitute of any inventive faculty themselves, do not readily conceive that others may possess it : They think of inventions as of miracles ; there might be such formerly, but they are ceased.

---

* Mr *F.* has since thought, that, possibly, the mutual repulsion of the inner opposite sides of the electrised cann, may prevent the accumulating an electric atmosphere upon them, and occasion it to stand chiefly on the outside. But recommends it to the farther examination of the curious.

With

Witl thefe, every une who offers a new invention is deem'd
a pretender: He had it from fome other country, or from
fome book: A man of *their own acquaintance* ; one who
has no more fenfe than themfelves, could not poffibly, in
their opinion, have been the inventer of any thing. They
are confirmed, too, in thefe fentiments, by frequent inftances of pretenfions to invention, which vanity is daily
producing. That vanity too, though an incitement to invention, is, at the fame time, the peft of inventors. Jealoufy and Envy deny the merit or the novelty of your invention ; but Vanity, when the novelty and merit are e-
ftablifhed, claims it for its own. The fmaller your invention is, the more mortification you receive in having the
credit of it difputed with you by a rival, whom the jealoufy and envy of others are ready to fupport againft you,
at leaft fo far as to make the point doubtful. It is not in
itfelf of importance enough for a difpute ; no one would
think your proofs and reafons worth their attention: And
yet if you do not difpute the point, and demonftrate your
right, you not only lofe the credit of being in that inftance
ingenious, but you fuffer the difgrace of not being *ingenu-*
ous ; not only of being a plagiary, but of being a plagiary
for trifles. Had the invention been greater it would have
difgrac'd you lefs ; for men have not fo contemptible an
idea of him that robs for gold on the highway, as of him
that can pick pockets for half-pence and farthings. Thus
through Envy, Jealoufy, and the Vanity of competitors for
Fame, the origin of many of the moft extraordinary inven-

tions,

tions, though produced within but a few centuries paft, is involved in doubt and uncertainty. We fcarce know to whom we are indebted for the *compafs*, and for *fpectacles*, nor have even *paper* and *printing*, that record every thing elfe, been able to preferve with certainty the name and reputation of their inventors. One would not, therefore, of all faculties, or qualities of the mind, wifh, for a friend, or a child, that he fhould have that of invention. For his attempts to benefit mankind in that way however well imagined, if they do not fucceed, expofe him, though very unjuftly, to general ridicule and contempt; and, if they do fucceed, to envy, robbery, and abufe.

I am, &c.

B. F.

LET-

LETTER XXV.

FROM

R. J. *Esq*; of *London*,

TO

BENJ. FRANKLIN, *Esq*; of *Philadelphia*.

DEAR SIR,

IT is now near three years since I received your excel‑
lent *Observations on the Increase of Mankind, &c**. in
which you have with so much sagacity and accuracy
shewn in what manner, and by what causes, that principal
means of political grandeur is best promoted ; and have
so well supported those just inferences you have occasional‑
ly drawn, concerning the general state of our *American*
colonies, and the views and conduct of some of the inhabi‑
tants of *Great Britain*.

You have abundantly proved that natural fecundity is
hardly to be considered, because the *vis generandi*, as far
as we know, is unlimited, and because experience shews
that the numbers of nations is altogether governed by col‑
lateral causes, and among these none of so much force as

* See page 197.

quan‑

quantity of fubfiftence, whether arifing from climate, foil, improvement of tillage, trade, fifheries, fecure property, conqueft of new countries, or other favourable circum-ftances.

As I perfectly concurred with you in your fentiments on thefe heads, I have been very defirous of building fome-what on the foundation you have there laid ; and was in-duced by your hints in the twenty-firft fection, to trouble you with fome thoughts on the influence manners have al-ways had, and are always likely to have on the numbers of a people, and their political profperity in general.

The end of every individual is its own private good. The rules it obferves in the purfuit of this good, are a fyf-tem of propofitions, almoft every one founded in authority, that is, derive their weight from the credit given to one or more perfons, and not from demonftration.

And this, in the moft important as well as the other af-fairs of life, is the cafe even of the wifeft and philofophical part of the human fpecies ; and that it fhould be fo is the lefs ftrange, when we confider that it is, perhaps, impoffi-ble to prove, that *being*, or life itfelf, has any other value than what is fet on it by authority.

A confirmation of this may be derived from the obfer-vation, that in every country in the univerfe, happinefs is fought upon a different plan ; and, even in the fame coun-try, we fee it placed by different ages, profeffions, and ranks of men, in the attainment of enjoyments utterly un-like.

Thefe

These propositions, as well as others, framed upon them, become habitual by degrees, and, as they govern the determination of the will, I call them *moral habits*.

There are another set of habits that have the direction of the members of the body, that I call therefore *mechanical habits*. These compose what we commonly call *The Arts*, which are more or less liberal or mechanical, as they more or less partake of assistance from the operations of the mind.

The *cumulus* of the moral habits of each individual, is the manners of that individual; the *cumulus* of the manners of individuals makes up the manners of a nation.

The happiness of individuals is evidently the ultimate end of political society; and political welfare, or the strength, splendour, and opulence of the state, have been always admitted, both by political writers, and the valuable part of mankind in general, to conduce to this end, and are therefore desirable.

The causes that advance or obstruct any one of these three objects, are external or internal. The latter may be divided into physical, civil, and personal, under which last head I comprehend the moral and mechanical habits of mankind. The physical causes are principally climate, soil, and number of subjects; the civil are government and laws; and political welfare is always in a ratio composed of the force of these particular causes; a multitude of external causes, and all these internal ones, not only controul and qualify, but are constantly acting on, and thereby in-

X x

fenfibly, as well as fenfibly, altering one another, both for the better and the worfe, and this not excepting the climate itfelf.

The powerful efficacy of manners in increafing a people, is manifeft from the inftance you mention, the Quakers; among them induftry and frugality multiplies and extends the ufe of the neceffaries of life; to manners of a like kind are owing the populoufnefs of *Holland, Switzerland, China, Japan* moft parts of *Indoftan, &c.* in every one of which the force of extent of territory and fertility of foil is multiplied, or their want compenfated by induftry and frugality.

Neither nature nor art have contributed much to the production of fubfiftence in *Switzerland,* yet we fee frugality preferves, and even increafes families that live on their fortunes, and which, in *England,* we call the Gentry; and the obfervation we cannot but make in the Southern part of this kingdom, that thofe families, including all fuperior ones, are gradually becoming extinct, affords the cleareft proof that luxury (that is, a greater expence of fubfiftence than in prudence a man ought to confume) is as deftructive as a proportionable want of it; but in *Scotland,* as in *Switzerland,* the Gentry, though one with another they have not one-fourth of the income, increafe in number.

And here I cannot help remarking, by the by, how well founded your diftinction is between the increafe of mankind in old and new fettled countries in general, and more particularly in the cafe of families of condition. In *America*

America, where their expences are more confined to neceffaries, and thofe neceffaries are cheap, it is common to
fee above one hundred perfons defcended from one living
old man. In *England* it frequently happens, where a man
has feven, eight, or more children, there has not been a
defcendant in the next generation, occafioned by the difficulties the number of children has brought on the family,
in a luxurious dear country, and which have prevented
their marrying.

That this is more owing to luxury than meer want, appears from what I have faid of *Scotland*, and more plainly
from parts of *England* remote from *London*, in moft of
which the neceffaries of life are nearly as dear, in fome dearer than in *London*, yet the people of all ranks marry and
breed up children.

Again ; among the lower ranks of life, none produce fo
few children as fervants. This is, in fome meafure, to be
attributed to their fituation, which hinders marriage, but
it is alfo to be attributed to their luxury, and corruption of
manners, which are greater than among any other fet of
people in *England*, and is the confequence of a nearer view
of the lives and perfons of a fuperior rank, than any inferi-
or rank, without a proper education, ought to have.

The quantity of fubfiftence in *England* has unqueftio-
nably become greater for many ages ; and yet if the inhabitants are more numerous, they certainly are not fo in
proportion to our improvement of the means of fupport.
I am apt to think there are few parts of this kingdom that

have

have not been at fome former time more populous than at prefent. I have feveral cogent reafons for thinking fo, of great part of the counties I am moft intimately acquainted with ; but as they were probably not all moft populous at the fame time, and as fome of our towns are vifibly and vaftly grown in bulk, I dare not fuppofe, as judicious men have done, that *England* is lefs peopled than heretofore.

This growth of our towns is the effect of a change of manners, and improvement of arts, common to all *Europe* ; and though it is not imagined that it has leffened the country growth of neceffaries, it has evidently, by introducing a greater confumption of them, (an infallible confequence of a nation's dwelling in towns) counteracted the effects of our prodigious advances in the arts.

But however frugality may fupply the place of, or prodigality counteract the effects of the natural or acquired fubfiftence of a country, induftry is, beyond doubt, a more efficacious caufe of plenty than any natural advantage of extent or fertility. I have mentioned inftances of frugality and induftry united with extent and fertility ; in *Spain* and *Afia* minor, we fee frugality joined to extent and fertility, without induftry ; in *Ireland* we once faw the fame ; *Scotland* had then none of them but frugality. The change in thefe two countries is obvious to every one, and it is owing to induftry not yet very widely diffufed in either.

The effects of induftry and frugality in *England* are furprizing ; both the rent and the value of the inheritance of land depend on them greatly more than on nature, and
this

this though there is no confiderable difference in the prices
of our markets. Land of equal goodnefs lets for double
the rent of other land lying in the fame county, and there
are many years purchafe difference between different counties, where rents are equally well paid and fecure.

Thus manners operate upon the number of inhabitants,
but of their filent effects upon a civil conftitution, hiftory
and even our own experience, yields us abundance of
proofs, though they are not uncommonly attributed to external caufes : Their fupport of a government againft external force is fo great, that it is a common maxim among
the advocates of liberty, that no free government was ever
diffolved, or overcome, before the manners of its fubjects
were corrupted.

The fuperiority of *Greece* over *Perfia*, was fingly owing
to their difference of manners ; and that, though all natural advantages were on the fide of the latter, to which I
might add the civil ones ; for though the greateft of all civil advantages, Liberty, was on the fide of *Greece*, yet that
added no political ftrength to her, than as it operated on her
manners, and, when they were corrupted, the reftoration
of their liberty by the *Romans*, overturned the remains of
their power.

Whether the manners of ancient *Rome* were, at any period, calculated to promote the happinefs of individuals, it
is not my defign to examine ; but that their manners, and
the effects of thofe manners on their government, and pub-

lick

lick conduct, founded, enlarged, and supported, and afterwards overthrew their empire, is beyond all doubt. One of the effects of their conquest furnishes us with a strong proof how prevalent manners are, even beyond quantity of subsistence ; for, when the custom of bestowing on the citizens of *Rome* corn enough to support themselves and families, was become established, and *Egypt* and *Sicily* produced the grain that fed the inhabitants of *Italy*, this became less populous every day, and the *Jus trium liberorum* was but an expedient that could not balance the want of industry and frugality.

But corruption of manners did not only thin the inhabitants of the *Roman* empire, it rendered the remainder incapable of defence, long before its fall, perhaps before the dissolution of the Republic : so that without standing disciplined armies composed of men, whose moral habits principally, and mechanical habits secondarily, made them different from the body of the people, the *Roman* empire had been a prey to the Barbarians many ages before it was.

By the mechanical habits of the soldiery, I mean their discipline, and the art of war ; and that this is but a secondary quality, appears from the inequality that has in all ages been between raw, though well disciplined armies, and veterans, and more from the irresistible force of a single moral habit, Religion, has conferred on troops frequently neither disciplined nor experienced.

The military manners of the Noblesse in *France*, compose the chief force of that kingdom, and the enterprizing

man-

manners, and reftlefs difpofitions of the inhabitants of *Ca-
nada* have enabled a handful of men to harrafs our populous, and, generally, lefs martial colonies ; yet neither are
of the value they feem at firft fight, becaufe, overbalanced
by the defect they occafion of other habits that would
produce more eligible political good : And military manners in a people are not neceffary in an age and country
where fuch manners may be occafionally formed and preferved among men enough to defend the ftate ; and fuch
a country is *Great-Britain*, where, though the lower clafs
of people are by no means of a military caft, yet they
make better foldiers than even the Noblefle of *France*.

The inhabitants of this country, a few ages back, were
to the populous and rich provinces of *France*, what *Cana-
da* is now to the *Britifh* colonies. It is true, there was
lefs difproportion between their natural ftrength ; but I
mean that the riches of *France* were a real weaknefs oppofed to the military manners founded upon poverty and a
rugged difpofition, then the character of the *Englifh* ; but
it muft be remembered, that at this time the manners of a
people were not diftinct from that of their foldiery, for
the ufe of ftanding armies has deprived a military people
of the advantages they before had over others ; and though
it has been often faid, that civil wars give power, becaufe
they render all men foldiers, I believe this has only been
found true in internal wars, following civil wars, and not
in external ones ; for now, in foreign wars, a fmall army
with ample means to fupport it, is of greater force than

one

one more numerous, with lefs. This laſt faƐt has often happened between *France* and *Germany*.

The means of fupporting armies, and, confequently, the power of exerting external ſtrength, are beſt found in the induſtry and frugality of the body of a people living under a government and laws that encourage commerce, for commerce is at this day almoſt the only *ſtimulus* that forces every one to contribute a ſhare of labour for the publick benefit.

But ſuch is the human frame, and the world is ſo conſtituted, that it is a hard matter to poſſeſs ones ſelf of a benefit, without laying ones ſelf open to a loſs on ſome other ſide ; the improvements of manners of one ſort, often deprave thoſe of another : Thus we ſee induſtry and frugality under the influence of commerce, which I call a commercial ſpirit, tend to deſtroy, as well as ſupport, the government it flouriſhes under.

Commerce perfeƐts the arts, but more the mechanical than the liberal, and this for an obvious reaſon ; it ſoftens and enervates the manners. Steady virtue, and unbending integrity, are ſeldom to be found where a ſpirit of commerce pervades every thing ; yet the perfeƐtion of commerce is, that every thing ſhould have its price. We every day ſee its progreſs, both to our benefit and detriment here. Things that *boni mores* are forbid to be ſet to ſale, are become its objeƐts, and there are few things indeed *extra commercium*. The legiſlative power itſelf has been *in commercio*, and church livings are ſeldom given without

con-

confideration, even by fincere Chriftians, and for confideration not feldom to very unworthy perfons. The rudenefs
of ancient military times, and the fury of more modern enthufiaftic ones, are worn off; even the fpirit of fo-
renfic contention is aftonifhingly diminifhed, all marks of
manners foftening; but Luxury and Corruption have taken
their places, and feem the infeparable companions of Commerce and the Arts.

I cannot help obferving, however, that this is much
more the cafe in extenfive countries, efpecially at their me-
tropolis, than in other places. It is an old obfervation of
politicians, and frequently made by hiftorians, that fmall
ftates always beft preferve their manners; whether this happens from the greater room there is for attention in the
legiflature, or from the lefs room there is for Ambition
and Avarice, it is a ftrong argument, among others, againft
an incorporating union of the colonies in *America*, or even
a federal one, that may tend to the future reducing them
under one government.

Their power, while united, is lefs, but their liberty, as
well as manners, is more fecure; and, confidering the little
danger of any conqueft to be made upon them, I had rather they fhould fuffer fomething through difunion, than
fee them under a general adminiftration lefs equitable than
that concerted at *Albany.*

I take it, the inhabitants of *Pennfylvania* are both frugal
and induftrious beyond thofe of any province in *America.*
If luxury fhould fpread, it cannot be extirpated by laws.

Y y We

We are told by *Plutarch*, that *Plato* ufed to fay, *It was a hard thing to make laws for the* Cyrenians, *a people abounding in plenty and opulence.*

But from what I fet out with, it is evident, if I be not miftaken, that education only can ftem the torrent, and without checking either true induftry or frugality, prevent the fordid frugality and lazinefs of the old *Irifh*, and many of the modern *Scotch*, (I mean the inhabitants of that country, thofe who leave it for another being generally induftrious) or the induftry, mixed with luxury, of this capital, from getting ground, and by rendering ancient manners familiar, produce a reconciliation between difintereftednefs and commerce ; a thing we often fee, but almoft always in men of a liberal education.

To conclude ; when we would form a people, foil and climate may be found, at leaft fufficiently good : Inhabitants may be encouraged to fettle, and even fupported for a while ; a good government ana laws may be framed, and even arts may be eftablifhed, or their produce imported ; but many neceffary moral habits are hardly ever found among thofe who voluntarily offer themfelves in times of quiet at home, to people new colonies; befides that the moral, as well as mechanical habits, adapted to a mother country, are frequently not fo to the new fettled one, and to external events, many of which are always unforefeen. Hence it is we have feen fuch fruitlefs attempts to fettle colonies, at an immenfe public and private expence, by feveral of the powers of *Europe* : And it is particularly obfer-

vable

vable that none of the *English* colonies became any way confiderable, till the neceffary manners were born and grew up in the county, excepting thofe to which fingular circumftances at home forced manners fit for the forming a new ftate.

I am, Sir, &c.   R. J.

L E T T E R XXVI.

F R O M

BENJAMIN FRANKLIN, *Efq*; of *Philadelphia,*

T O

Dr *L——*, at *Charles-Town, South-Carolina.*

S I R,   *New-York, April* 14, 1757.

IT is a long time fince I had the pleafure of a line from you ; and, indeed, the troubles of our country, with the hurry of bufinefs I have been engaged in on that account, have made me fo bad a correfpondent, that I ought not to expect punctuality in others.

But being about to embark for *England,* I could not quit the Continent without paying my refpects to you, and, at the fame time, taking leave to introduce to your acquaintance a Gentleman of learning and merit, Colonel

Y y 2   *Henry*

Henry Bouquet, who does me the favour to prefent you this letter, and with whom I am fure you will be much pleafed.

Profeffor *Simpfon*, of *Glafgow*, lately communicated to me fome curious experiments of a phyfician of his acquaintance, by which it appeared, that an extraordinary degree of cold, even to freezing, might be produced by evaporation. I have not had leifure to repeat and examine more than the firft and eafieft of them, *viz.*—Wet the ball of a thermometer by a feather dipt in fpirit of wine, which has been kept in the fame room, and has, of courfe, the fame degree of heat or cold. The mercury finks prefently three or four degrees, and the quicker, if, during the evaporation, you blow on the ball with bellows; a fecond wetting and blowing, when the mercury is down, carries it yet lower. I think I did not get it lower than five or fix degrees from where it naturally ftood, which was, at that time, fixty. But it is faid, that a veffel of water being placed in another fomewhat larger, containing fpirit, in fuch a manner that the veffel of water is furrounded with the fpirit, and both placed under the receiver of an air-pump; on exhaufting the air, the fpirit, evaporating, leaves fuch a degree of cold as to freeze the water, though the thermometer, in the open air, ftands many degrees above the freezing point.

I know not how this phenomenon is to be accounted for, but it gives me occafion to mention fome loofe notions relating to heat and cold, which I have for fome time

enter-

entertained, but not yet reduced into any form. Allowing common fire, as well as electrical, to be a fluid capable of permeating other bodies, and feeking an equilibrium, I imagine fome bodies are better fitted by nature to be conductors of that fluid than others ; and that, generally, thofe which are the beft conductors of the electrical fluid, are alfo the beft conductors of this ; and *e contra.*

Thus a body which is a good conductor of fire, readily receives it into its fubftance, and conducts it through the whole to all the parts, as metals and water do ; and if two bodies, both good conductors, one heated, the other in its common ftate, are brought into contact with each other, the body which has moft fire, readily communicates of it to that which had leaft, and that which had leaft readily receives it, till an equilibrium is produced. Thus if you take a dollar between your fingers with one hand, and a piece of wood, of the fame dimenfions, with the other, and bring both at the fame time to the flame of a candle, you will find yourfelf obliged to drop the dollar before you drop the wood, becaufe it conducts the heat of the candle fooner to your flefh. Thus if a filver tea-pot had a handle of the fame metal, it would conduct the heat from the water to the hand, and become too hot to be ufed ; we therefore give to a metal tea-pot a handle of wood, which is not fo good a conductor as metal. But a china or ftone tea-pot being in fome degree of the nature of glafs, which is not a good conductor of heat, may have a handle of the fame ftuff. Thus, alfo, a damp moift air fhall make

a man

a man more senfible of cold, or chill him more than a dry
air that is colder, becaufe a moift air is fitter to receive and
conduct away the heat of his body. This fluid entering
bodies in great quantity, firft expands them, by feparating
their parts a little, afterwards by farther feparating their
parts, it renders folids fluid, and at length diffipates their
parts in air. Take this fluid from melted lead, or from
water, the parts cohere again, the firft grows folid, the latter becomes ice : And this is fooner done by the means of
good conductors. Thus if you take, as I have done, a
fquare bar of lead, four inches long, and one inch thick,
together with three pieces of wood planed to the fame
dimenfions, and lay them, as in the margin,
on a fmooth board, fixt fo as not to be eafily
feparated or moved, and pour into the cavity
they form, as much melted lead as will fill
it, you will fee the melted lead chill, and
become firm, on the fide next the leaden
bar, fome time before it chills on the other three fides
in contact with the wooden bars, though before the lead
was poured in, they might all be fuppofed to have the
fame degree of heat or coldnefs, as they had been expofed
in the fame room to the fame air. You will likewife obferve, that the leaden bar, as it has cooled the melted lead
more than the wooden bars have done, fo it is itfelf more
heated by the melted lead. There is a certain quantity of
this fluid called fire, in every living human body, which
fluid being in due proportion, keeps the parts of the flefh
and

and blood, at fuch a juft diftance from each other, as that
the flefh and nerves are fupple, and the blood fit for circulation. If part of this due proportion of fire be conducted away by means of a contact with other bodies, as
air, water, or metals, the parts of our fkin and flefh that
come into fuch contact, firft draw more near together than
is agreeable, and give that fenfation which we call cold ;
and if too much be conveyed away, the body ftiffens, the
blood ceafes to flow, and death enfues. On the other
hand, if too much of this fluid be communicated to the
flefh, the parts are feparated too far, and pain enfues, as
when they are feparated by a pin or lancet. The fenfation
that the feparation by fire occafions, we call heat, or burning. My defk on which I now write, and the lock of my
defk, are both expofed to the fame temperature of the air,
and have therefore the fame degree of heat or cold ; yet if
I lay my hand fucceffively on the wood and on the metal,
the latter feels much the coldeft, not that it is really fo, but
being a better conductor, it more readily than the wood
takes away and draws into itfelf the fire that was in my
fkin. Accordingly if I lay one hand, part on the lock,
and part on the wood, and after it had lain fo fome time,
I feel both parts with my other hand, I find the part that
has been in contact with the lock, very fenfibly colder to
the touch, than the part that lay on the wood. How a
living animal obtains its quantity of this fluid called fire, is a
curious queftion. I have fhewn that fome bodies (as metals) have a power of attracting it ftronger than others ; and

I have

I have fometimes fufpected that a living body had fome power of attracting out of the air, or other bodies, the heat it wanted. Thus metals hammered, or repeatedly bent, grow hot in the bent or hammered part. But when I confider that air, in contact with the body, cools it; that the furrounding air is rather heated by its contact with the body; that every breath of cooler air drawn in, carries off part of the body's heat when it paffes out again; that therefore there muft be in the body a fund for producing it, or otherwife the animal would foon grow cold; I have been rather inclined to think that the fluid *fire*, as well as the fluid *air*, is attracted by plants in their growth, and be-comes confolidated with the other materials of which they are formed, and makes a great part of their fubftance : That when they come to be digefted, and to fuffer in the veffels a kind of fermentation, part of the fire, as well as part of the air, recovers its fluid active ftate again, and dif-fufes itfelf in the body digefting and feparating it : That the fire fo reproduced, by digeftion and feparation, continually leaving the body, its place is fupplied by frefh quantities, arifing from the continual feparation. That whatever quickens the motion of the fluids in an animal, quickens the feparation, and reproduces more of the fire; as exercife. That all the fire emitted by wood, and other combuftibles, when burning, exifted in them before, in a folid ftate, be-ing only difcovered when feparating. That fome foffils, as fulphur, fea-coal, &c. contain a great deal of folid fire : That fome bodies are almoft wholly folid fire; and that,

in

in fhort, what efcapes and is diffipated in the burning of bodies, befides water and earth, is generally the air and fire that before made parts of the folid.—Thus I imagine that animal heat arifes by or from a kind of fermentation in the juices of the body, in the fame manner as heat arifes in the liquors preparing for diftillation, wherein there is a fepa-ration of the fpirituous, from the watry and earthy parts. —And it is remarkable, that the liquor in a diftiller's vat, when in its higheft and beft ftate of fermentation, as I have been informed, has the fame degree of heat with the hu-man body; that is, about 94 or 96.

Thus, as by a conftant fupply of fuel in a chimney, you keep a warm room, fo, by a conftant fupply of food in the ftomach, you keep a warm body; only where little exer-cife is ufed, the heat may poffibly be conducted away too faft; in which cafe fuch materials are to be ufed for cloath-ing and bedding, againft the effects of an immediate con-tact of the air, as are, in themfelves, bad conductors of heat, and, confequently, prevent its being communicated thro' their fubftance to the air. Hence what is called *warmth* in wool, and its preference, on that account, to linnen; wool not being fo good a conductor: And hence all the natural coverings of animals, to keep them warm, are fuch as retain and confine the natural heat in the body, by being bad conductors, fuch as wool, hair, feathers, and the filk by which the filk-worm, in its tender embrio ftate, is firft cloathed. Cloathing, thus confidered, does not make a man warm by *giving* warmth, but by *preventing* the

too

too quick diffipation of the heat produced in his body, and fo occafioning an accumulation.

There is another curious queftion I will juft venture to touch upon, *viz.* Whence arifes the fudden extraordinary degree of cold, perceptible on mixing fome chemical liquors, and even on mixing falt and fnow, where the compofition appears colder than the coldeft of the ingredients ? I have never feen the chemical mixtures made, but falt and fnow I have often mixed myfelf, and am fully fatisfied that the compofition feels much colder to the touch, and lowers the mercury in the thermometer more than either ingredient would do feparately. I fuppofe, with others, that cold is nothing more than the abfence of heat or fire. Now if the quantity of fire before contained or diffufed in the fnow and falt, was expelled in the uniting of the two matters, it muft be driven away either through the air or the veffel containing them. If it is driven off through the air, it muft warm the air, and a thermometer held over the mixture, without touching it, would difcover the heat, by the rifing of the mercury, as it muft, and always does in warm air.

This, indeed, I have not tried, but I fhould guefs it would rather be driven off through the veffel, efpecially if the veffel be metal, as being a better conductor than air; and fo one fhould find the bafon warmer after fuch mixture. But, on the contrary, the veffel grows cold, and even water in which the veffel is fometimes placed for the experiment, freezes into hard ice on the bafon. Now I know

not

not how to account for this, otherwife than by fuppofing that the compofition is a better conductor of fire than the ingredients feparately, and, like the lock compared with the wood, has a ftronger power of attracting fire, and does accordingly attract it fuddenly from the fingers, or a thermometer put into it, from the bafon that contains it, and from the water in contact with the outfide of the bafon; fo that the fingers have the fenfation of extreme cold, by being deprived of much of their natural fire; the thermometer finks, by having part of its fire drawn out of the mercury; the bafon grows colder to the touch, as by having its fire drawn into the mixture, it is become more capable of drawing and receiving it from the hand; and through the bafon, the water lofes its fire that kept it fluid; fo it becomes ice.—One would expect that from all this attracted acquifition of fire to the compofition, it fhould become warmer; and, in fact, the fnow and falt diffolve at the fame time into water, without freezing.

I am, Sir, &c. B. F.

LETTER XXVII.

FROM

Benjamin Franklin, *Esq*; of *Philadelphia*,

T O

Peter Collinson, *Esq*; at *London*.

S I R,

ACCORDING to your requeſt, I now ſend you the Arithmetical Curioſity, of which this is the hiſtory.

Being one day in the country, at the houſe of our common friend, the late learned Mr. *Logan*, he ſhewed me a folio *French* book, filled with magic ſquares, wrote, if I forget not, by one M. *Frenicle*, in which he ſaid the author had diſcovered great ingenuity and dexterity in the management of numbers; and, though ſeveral other foreigners had diſtinguiſhed themſelves in the ſame way, he did not recollect that any one *Engliſhman* had done any thing of the kind remarkable.

I ſaid, it was, perhaps, a mark of the good ſenſe of our *Engliſh* mathematicians, that they would not ſpend their time in things that were merely *difficiles nugæ*, incapable of any uſeful application. He anſwered, that many of the arithmetical or mathematical queſtions, publickly propoſed

and

and anfwered in *England*, were equally trifling and ufelefs. Perhaps the confidering and anfwering fuch queftions, I replied, may not be altogether ufelefs, if it produces by practice an habitual readinefs and exactnefs in mathematical difquifitions, which readinefs may, on many occafions, be of real ufe. In the fame way, fays he, may the making of thefe fquares be of ufe. I then confeffed to him, that in my younger days, having once fome leifure, (which I ftill think I might have employed more ufefully) I had amufed myfelf in making thefe kind of magic fquares, and, at length, had acquired fuch a knack at it, that I could fill the cells of any magic fquare, of reafonable fize, with a feries of numbers as faft as I could write them, difpofed in fuch a manner, as that the fums of every row, horizontal, perpendicular, or diagonal, fhould be equal; but not being fatisfied with thefe, which I looked on as common and eafy things, I had impofed on myfelf more difficult tafks, and fucceeded in making other magic fquares, with a variety of properties, and much more curious. He then fhewed me feveral in the fame book, of an uncommon and more curious kind; but as I thought none of them equal to fome I remembered to have made, he defired me to let him fee them; and accordingly, the next time I vifited him, I carried him a fquare of 8, which I found among my old papers, and which I will now give you, with an account of its properties. *(See Plate IV.)*

The

The properties are,

1. That every ſtrait row (horizontal or vertical) of 8 numbers added together, makes 260, and half each row half 260.

2. That the bent row of 8 numbers, aſcending and deſcending diagonally, *viz.* from 16 aſcending to 10, and from 23 deſcending to 17 ; and every one of its parallel bent rows of 8 numbers, make 260.—Alſo the bent row from 52, deſcending to 54, and from 43 aſcending to 45 ; and every one of its parallel bent rows of 8 numbers, make 260.—Alſo the bent row from 45 to 43 deſcending to the left, and from 23 to 17 deſcending to the right, and every one of its parallel bent rows of 8 numbers make 260.— Alſo the bent row from 52 to 54 deſcending to the right, and from 10 to 16 deſcending to the left, and every one of its parallel bent rows of 8 numbers make 260.—Alſo the parallel bent rows next to the above-mentioned, which are ſhortened to 3 numbers aſcending, and 3 deſcending, *&c.* as from 53 to 4 aſcending, and from 29 to 44 deſcending, make, with the 2 corner numbers, 260.—Alſo the 2 numbers 14, 61 aſcending, and 36, 19 deſcending, with the lower 4 numbers ſituated like them, *viz.* 50, 1, deſcending, and 32, 47, aſcending, make 260.—And, laſtly, the 4 corner numbers, with the 4 middle numbers, make 260.

So this magical ſquare ſeems perfect in its kind. But theſe are not all its properties ; there are 5 other curious ones, which, at ſome other time, I will explain to you.

Mr.

PLATE IV.

A Magic Square of Squares.

200	217	232	249	8	25	40	57	72	89	104	121	136	153	168	181
58	39	26	7	250	231	218	199	186	167	154	135	122	103	90	71
198	219	230	251	6	27	38	59	70	91	102	123	134	155	166	187
60	37	28	5	252	229	220	197	188	165	156	133	124	101	92	69
201	216	233	248	9	24	41	56	73	88	105	120	137	152	169	184
55	42	23	10	247	234	215	202	183	170	151	138	119	106	87	74
203	214	235	246	11	22	43	54	75	86	107	118	139	150	171	182
53	44	21	12	245	236	213	204	181	172	149	140	117	108	85	76
205	212	237	244	13	20	45	52	77	84	109	116	141	148	173	180
51	46	19	14	243	238	211	206	179	174	147	142	115	110	83	78
207	210	239	242	15	18	47	50	79	82	111	114	143	146	175	178
49	48	17	16	241	240	209	208	177	176	145	144	113	112	81	80
196	221	228	253	4	29	36	61	68	93	100	125	132	157	164	189
62	35	30	3	254	227	222	195	190	163	158	131	126	99	94	67
194	223	226	255	2	31	34	63	66	95	98	127	130	159	162	191
64	33	32	1	256	225	224	193	192	161	160	129	128	97	96	65

Mr. *Logan* then fhewed me an old arithmetical book, in quarto, wrote, I think, by one *Stifelius*, which contained a fquare of 16, that he faid he fhould imagine muft have been a work of great labour; but if I forget not, it had only the common properties of making the fame fum, *viz.* 2056,, in every row, horizontal, vertical, and diagonal. Not willing to be out-done by Mr *Stifelius*, even in the fize of my fquare, I went home, and made, that evening, the following magical fquare of 16, which, befides having all the properties of the foregoing fquare of 8, *i. e.* it would make the 2056 in all the fame rows and diagonals, had this added, that a four fquare hole being cut in a piece of paper of fuch a fize as to take in and fhew through it, juft 16 of the little fquares, when laid on the greater fquare, the fum of the 16 numbers fo appearing through the hole, wherever it was placed on the greater fquare, fhould likewife make 2056. This I fent to our friend the next morning, who, after fome days, fent it back in a letter, with thefe words :——" I return to thee thy afto-" nifhing or moft ftupendous piece of the magical fquare, " in which" ——— but the compliment is too extravagant, and therefore, for his fake, as well as my own, I ought not to repeat it. Nor is it neceffary ; for I make no queftion but you will readily allow this fquare of 16 to be the moft magically magical of any magic fquare ever made by any magician. (*See the Plate.*)

I did

I did not, however, end with fquares, but compofed alfo a magick circle, confifting of 8 concentric circles, and
8 radial rows, filled with a feries of numbers, from 12 to
75, inclufive, fo difpofed as that the numbers of each circle,
or each radial row, being added to the central number 12,
they made exactly 360, the number of degrees in a circle;
and this circle had, moreover, all the properties of the
fquare of 8.   If you defire it, I will fend it; but at prefent,
I believe, you have enough on this fubject.

I am, &c.   B. F.

L E T T E R XXVIII.

To the fame.

S I R,

I AM glad the perufal of the magical fquares afforded
you any amufement.   I now fend you the magical
circle.   (*See Plate* V.)

Its properties, befides thofe mentioned in my former,
are thefe.

Half the number in any radial row, added with half the
central number, make 280, equal to the number of degrees
in a femi-circle.

Alfo

PLATE V. *A Magic Circle of Circles.* *Page.* 35.

Alfo half the numbers in any one of the concentric circles, taken either above or below the horizontal double
line, with half the central number, make 180.

And if any four adjoining numbers, ftanding nearly in a
fquare, be taken from any part, and added with half the
central number, they make 180.

There are, moreover, included four other fets of circular fpaces, excentric with refpect to the firft, each of thefe
fets containing five fpaces. The centers of the circles
that bound them, are at A, B, C, and D. Each fet, for
the more eafy diftinguifhing them from the firft, are drawn
with a different colour'd ink, red, blue, green, and yellow*

Thefe fets of excentric circular fpaces interfect thofe of
the concentric, and each other ; and yet the numbers contained in each of the twenty excentric fpaces, taken all a-
round, make, with the central number, the fame fum as
thofe in each of the 8 concentric, *viz.* 360. The halves,
alfo of thofe drawn from the centers A and C, taken above
or below the double horizontal line, and of thofe drawn
from centers B and D, taken to the right or left of the vertical line, do, with half the central number, make juft 180.

It may be obferved, that there is not one of the numbers
but what belongs at leaft to two of the different circular
fpaces; fome to three, fome to four, fome to five ; and yet
they are all fo placed as never to break the required num-

* In the plate they are diftinguifhed by dafhed or dotted lines, as different as the engraver could well make them.

ber

ber 360, in any of the 28 circular spaces within the primitive circle.

These interwoven circles make so perplexed an appearance, that it is not easy for the eye to trace every circle of numbers one would examine, through all the maze of circles intersected by it; but if you fix one foot of the compasses in either of the centers, and extend the other to any number in the circle you would examine belonging to that center, the moving foot will point the others out, by passing round over all the numbers of that circle successively.

I am, &c. B. F.

————————————————

LETTER XXIX.

To the same.

Dear Sir, *Philadelphia, Aug.* 25, 1755.

AS you have my former papers on Whirlwinds, &c. I now send you an account of one which I had lately an opportunity of seeing and examining myself.

Being in *Maryland*, riding with Col. *Tasker*, and some other gentlemen to his country-seat, where I and my son were entertained by that amiable and worthy man, with great hospitality and kindness, we saw in the vale below us, a small whirlwind beginning in the road, and shewing itself by the dust it raised and contained. It appeared in

the

the form of a sugar-loaf, spinning on its point, moving up the hill towards us, and enlarging as it came forward. When it passed by us, its smaller part near the ground, appeared not bigger than a common barrel, but widening upwards, it seemed, at 40 or 50 feet high, to be 20 or 30 feet in diameter. The rest of the company stood looking after it, but my curiosity being stronger, I followed it, riding close by its side, and observed its licking up, in its progress, all the dust that was under its smaller part. As it is a common opinion that a shot, fired through a water-spout, will break it, I tried to break this little whirl-wind, by striking my whip frequently through it, but without any effect. Soon after, it quitted the road and took into the woods, growing every moment larger and stronger, raising, instead of dust, the old dry leaves with which the ground was thick covered, and making a great noise with them and the branches of the trees, bending some tall trees round in a circle swiftly and very surprizingly, though the progressive motion of the whirl was not so swift but that a man on foot might have kept pace with it, but the circular motion was amazingly rapid. By the leaves it was now filled with, I could plainly perceive that the current of air they were driven by, moved upwards in a spiral line ; and when I saw the trunks and bodies of large trees invelop'd in the passing whirl, which continued intire after it had left them, I no longer wondered that my whip had no effect on it in its smaller state. I accom-

panied

panied it about three quarters of a mile, till some limbs of
dead trees, broken off by the whirl, flying about, and
falling near me, made me more apprehensive of danger ;
and then I stopped, looking at the top of it as it went on,
which was visible, by means of the leaves contained in it,
for a very great height above the trees. Many of the leaves,
as they got loose from the upper and widest part, were scat-
tered in the wind ; but so great was their height in the air,
that they appeared no bigger than flies. My son, who
was, by this time, come up with me, followed the whirlwind till it left the woods, and crossed an old tobacco-field,
where, finding neither dust nor leaves to take up, it gradually became invisible below as it went away over that field.
The course of the general wind then blowing was along
with us as we travelled, and the progressive motion of the
whirlwind was in a direction nearly opposite, though it
did not keep a strait line, nor was its progressive motion
uniform, it making little sallies on either hand as it went,
proceeding sometimes faster, and sometimes slower, and
seeming sometimes for a few seconds almost stationary,
then starting forwards pretty fast again. When we re-
joined the company, they were admiring the vast height
of the leaves, now brought by the common wind, over our
heads. These leaves accompanied us as we travelled,
some falling now and then round about us, and some not
reaching the ground till we had gone near three miles from
the place where we first saw the whirlwind begin. Upon

my

my asking Col. *Tasker* if such whirlwinds were common in *Maryland*, he answered pleasantly, *No, not at all common ; but we got this on purpose to treat Mr.* Franklin. And a very high treat it was, to

Dear Sir,
Your affectionate friend,
and humble servant

B. F.

L E T T E R   XXX.

T O

J OHN P RINGLE, M.D. and F.R.S.

S I R,　　　　　　*Craven-street, Dec.* 21, 1757.

IN compliance with your request, I send you the following account of what I can at present recollect relating to the effects of electricity in paralytic cases, which have fallen under my observation.

Some years since, when the news-papers made mention of great cures performed in *Italy* and *Germany*, by means of electricity, a number of paralytics were brought to me from different parts of *Pensylvania*, and the neighbouring provinces, to be electrised, which I did for them at their request. My method was, to place the patient first in a

chair,

chair, on an electric ftool, and draw a number of large ftrong fparks from all parts of the affected limb or fide. Then I fully charged two fix-gallon glafs jars, each of which had about three fquare feet of furface coated; and I fent the united fhock of thefe through the affected limb or limbs, repeating the ftroke commonly three times each day. The firft thing obferved, was an immediate great- er fenfible warmth in the lame limbs that had received the ftroke, than in the others; and the next morning the pa- tients ufually related, that they had in the night felt a pricking fenfation in the flefh of the paralytic limbs; and would fometimes fhew a number of fmall red fpots, which they fuppofed were occafioned by thofe prickings. The limbs, too, were found more capable of voluntary motion, and feemed to receive ftrength. A man, for inftance, who could not the firft day lift the lame hand from off his knee, would the next day raife it four or five inches, the third day higher; and on the fifth day was able, but with a feeble languid motion, to take off his hat. Thefe appear- ances gave great fpirits to the patients, and made them hope a perfect cure; but I do not remember that I ever faw a- ny amendment after the fifth day; which the patients per- ceiving, and finding the fhocks pretty fevere, they became difcouraged, went home, and in a fhort time relapfed; fo that I never knew any advantage from electricity in palfies that was permanent. And how far the apparent temporary advantage might arife from the exercife in the

patients

patients journey, and coming daily to my houfe, or from the fpirits given by the hope of fuccefs, enabling them to exert more ftrength in moving their limbs, I will not pretend to fay.

Perhaps fome permanent advantage might have been obtained, if the electric fhocks had been accompanied with proper medicine and regimen, under the direction of a fkilful phyfician. It may be, too, that a few great ftrokes as given in my method, may not be fo proper as many fmall ones; fince, by the account from *Scotland*, of a cafe, in which two hundred fhocks from a phial were given daily, it feems, that a perfect cure has been made. As to any uncommon ftrength fuppofed to be in the machine ufed in that cafe, I imagine it could have no fhare in the effect produced; fince the ftrength of the fhock from charged glafs, is in proportion to the quantity of furface of the glafs coated; fo that my fhocks from thofe large jars, muft have been much greater than any that could be received from a phial held in the hand.

I am, with great refpect,

S I R,

Your moft obedient Servant,

B. F.

L E T-

LETTER XXXI.

To the fame.

S I R,   *Craven-ftreet, Jan.* 6, 1758.

I Return Mr. *Mitchell*'s paper on the ftrata of the earth* with thanks. The reading of it, and perufal of the draft that accompanies it, have reconciled me to thofe convulfions which all naturalifts agree this globe has fuffered. Had the different ftrata of clay, gravel, marble, coals, lime-ftone, fand, minerals, &c. continued to lie level, one under the other, as they may be fuppofed to have done before thofe convulfions, we fhould have had the ufe only of a few of the uppermoft of the ftrata, the others lying too deep and too difficult to be come at ; but the fhell of the earth being broke, and the fragments thrown into this oblique pofition, the disjointed ends of a great number of ftrata of different kinds are brought up to day, and a great variety of ufeful materials put into our power, which would otherwife have remained eternally concealed from us. So that what has been ufually looked upon as a *ruin* fuffered by this part of the univerfe, was, in reality, only a preparation, or means of rendering the earth more fit for ufe, more capable of being to mankind a convenient and comfortable habitation.

I am, Sir, with great efteem, yours, &c.   B. F.

* See this Paper afterwards printed in the *Philofophical Tranfactions.*

L E T T E R XXXII.

To Dr. *L.* of *Charles-Town, South-Carolina.*

Dear Sir,   *London, June* 17, 1758.

IN a former letter I mentioned the experiment for cooling bodies by evaporation, and that I had, by repeatedly wetting the thermometer with common fpirits, brought the mercury down five or fix degrees. Being lately at *Cambridge,* and mentioning this in converfation with Dr. *Hadley,* profeffor of chemiftry there, he propofed repeating the experiments with ether, inftead of common fpirits, as the ether is much quicker in evaporation. We accordingly went to his chamber, where he had both ether and a thermometer. By dipping firft the ball of the thermometer into the ether, it appeared that the ether was precifely of the fame temperament with the thermometer, which ftood then at 65; for it made no alteration in the height of the little column of mercury. But when the thermometer was taken out of the ether, and the ether with which the ball was wet, began to evaporate, the mercury funk feveral degrees. The wetting was then repeated by a feather that had been dipped into the ether, when the mercury funk ftill lower. We continued this operation, one of us wetting the ball, and another of the company

B b b

blowing

blowing on it with the bellows, to quicken the evaporation, the mercury finking all the time, till it came down to
7, which is 25 degrees below the freezing point, when we
left off.—Soon after it paffed the freezing point, a thin coat
of ice began to cover the ball. Whether this was water
collected and condenfed by the coldnefs of the ball, from
the moifture in the air, or from our breath ; or whether
the feather, when dipped into the ether, might not fometimes go through it, and bring up fome of the water that
was under it, I am not certain ; perhaps all might contribute. The ice continued increafing till we ended the
experiment, when it appeared near a quarter of an inch
thick all over the ball, with a number of fmall fpicula,
pointing outwards. From this experiment one may fee
the poffibility of freezing a man to death on a warm fummer's day, if he were to ftand in a paffage thro which
the wind blew brifkly, and to be wet frequently with e-
ther, a fpirit that is more inflammable than brandy, or
common fpirits of wine.

It is but within thefe few years, that the *European* phi-
lofophers feem to have known this power in nature, of
cooling bodies by evaporation. But in the eaft they have
long been acquainted with it. A friend tells me, there is
a paffage in *Bernier*'s travels through *Indoftan*, written near
one hundred years ago, that mentions it as a practice (in
travelling over dry defarts in that hot climate) to carry water in flafks wrapt in wet woollen cloths, and hung on the

fhady

fhady fide of the camel, or carriage, but in the free air; whereby, as the cloths gradually grow drier, the water contained in the flafks is made cool. They have likewife a kind of earthen pots, unglaz'd, which let the water gradually and flowly ooze through their pores, fo as to keep the outfide a little wet, notwithftanding the continual evaporation, which gives great coldnefs to the veffel, and the water contained in it. Even our common failors feem to have had fome notion of this property; for I remember, that being at fea, when I was a youth, I obferved one of the failors, during a calm in the night, often wetting his finger in his mouth, and then holding it up in the air, to difcover, as he faid, if the air had any motion, and from which fide it came; and this he expected to do, by finding one fide of his finger grow fuddenly cold, and from that fide he fhould look for the next wind; which I then laughed at as a fancy.

May not feveral phænomena, hitherto unconfidered, or unaccounted for, be explained by this property? During the hot *Sunday* at *Philadelphia,* in *June* 1750, when the thermometer was up at 100 in the fhade, I fat in my chamber without exercife, only reading or writing, with no other cloaths on than a fhirt, and a pair of long linen drawers, the windows all open, and a brifk wind blowing through the houfe, the fweat ran off the backs of my hands, and my fhirt was often fo wet, as to induce me to call for dry ones to put on; in this fituation, one might

have

have expected, that the natural heat of the body 96, added to the heat of the air 100, should jointly have created or produced a much greater degree of heat in the body ; but the fact was, that my body never grew so hot as the air that surrounded it, or the inanimate bodies immers'd in the same air. For I remember well, that the desk, when I laid my arm upon it ; a chair, when I sat down in it ; and a dry shirt out of the drawer, when I put it on, all felt exceeding warm to me, as if they had been warmed before a fire. And I suppose a dead body would have acquired the temperature of the air, though a living one, by continual sweating, and by the evaporation of that sweat, was kept cold.—May not this be a reason why our reapers in *Pensylvania*, working in the open field, in the clear hot sunshine common in our harvest-time *, find themselves well able to go through that labour, without being much incommoded by the heat, while they continue to sweat, and while they supply matter for keeping up that sweat, by drinking frequently of a thin evaporable liquor, water mixed with rum ; but if the sweat stops, they drop, and sometimes die suddenly, if a sweating is not again brought on by drinking that liquor, or, as some rather chuse in that case, a kind of hot punch, made with water, mixed with

* *Pensylvania* is in about lat. 40, and the sun, of course, about 12 degrees higher, and therefore much hotter than in *England*. Their harvest is about the end of *June*, or beginning of *July*, when the sun is nearly at the highest.

honey,

honey, and a confiderable proportion of vinegar ?—May there not be in negroes a quicker evaporation of the perfpi-rable matter from their fkins and lungs, which, by cool-ing them more, enables them to bear the fun's heat better than whites do ? (if that is a fact, as it is faid to be ; for the alledg'd neceffity of having negroes rather than whites, to work in the *Weft-India* fields, is founded upon it) though the colour of their fkins would otherwife make them more fenfible of the fun's heat, fince black cloth heats much fooner, and more, in the fun, than white cloth. I am perfuaded, from feveral inftances happening within my knowledge, that they do not bear cold weather fo well as the whites ; they will perifh when expofed to a lefs de-gree of it, and are more apt to have their limbs froft-bitten ; and may not this be from the fame caufe ?   Would not the earth grow much hotter under the fummer fun, if a con-ftant evaporation from its furface, greater as the fun fhines ftronger, did not, by tending to cool it, balance, in fome degree, the warmer effects of the fun's rays ?—Is it not owing to the conftant evaporation from the furface of e-very leaf, that trees, though fhone on by the fun, are al-ways, even the leaves themfelves, cool to our fenfe ? at leaft much cooler than they would otherwife be ?—May it not be owing to this, that fanning ourfelves when warm, does really cool us, though the air is itfelf warm that we drive with the fan upon our faces ; for the atmofphere round, and next to our bodies, having imbibed as much of the

perfpired

perfpired vapour as it can well contain, receives no more, and the evaporation is therefore check'd and retarded, till we drive away that atmofphere, and bring dryer air in its place, that will receive the vapour, and thereby facilitate and increafe the evaporation ? Certain it is, that mere blowing of air on a dry body does not cool it, as any one may fatisfy himfelf, by blowing with a bellows on the dry ball of a thermometer; the mercury will not fall; if it moves at all, it rather rifes, as being warmed by the friction of the air on its furface ?—To thefe queries of i-magination, I will only add one practical obfervation; that wherever it is thought proper to give eafe, in cafes of painful inflammation in the flefh, (as from burnings, or the like) by cooling the part ; linen cloths, wet with fpirit, and applied to the part inflamed, will produce the coolnefs re-quired, better than if wet with water, and will continue it longer. For water, though cold when firft applied, will foon acquire warmth from the flefh, as it does not evaporate faft enough ; but the cloths wet with fpirit, will continue cold as long as any fpirit is left to keep up the evaporation, the parts warmed efcaping as foon as they are warmed, and carrying off the heat with them.

I am, Sir, &c.   B. F.

LET-

L E T T E R XXXIII.

T O

J. B. Efq; *at* Bofton, *in* New-England.

Dear Sir,   *London, Dec.* 2, 1758.

I HAVE executed here an eafy fimple contrivance, that I have long fince had in fpeculation, for keeping rooms warmer in cold weather than they generally are, and with lefs fire. It is this. The opening of the chimney is con-tracted, by brick-work faced with marble flabs, to about two feet between the jambs, and the breaft brought down to within about three feet of the hearth.—An iron frame is placed juft under the breaft, and extending quite to the back of the chimney, fo that a plate of the fame metal may flide horizontally backwards and forwards in the grooves on each fide of the frame. This plate is juft fo large as to fill the whole fpace, and fhut the chimney en-tirely when thruft quite in, which is convenient when there is no fire; drawing it out, fo as to leave a fpace between its farther edge and the back, of about two inches; this fpace is fufficient for the fmoke to pafs; and fo large a part of

the

the funnel being ftopt by the reft of the plate, the paffage of warm air out of the room, up the chimney, is obftructed and retarded, and by that means much cold air is prevented from coming in through crevices, to fupply its place. This effect is made manifeft three ways. Firft, when the fire burns brifkly in cold weather, the howling or whiftling noife made by the wind, as it enters the room through the crevices, when the chimney is open as ufual, ceafes as foon as the plate is flid in to its proper diftance. Secondly, opening the door of the room about half an inch, and holding your hand againft the opening, near the top of the door, you feel the cold air coming in againft your hand, but weakly, if the plate be in. Let another perfon fuddenly draw it out, fo as to let the air of the room go up the chimney, with its ufual freedom where chimneys are open, and you immediately feel the cold air rufhing in ftrongly. Thirdly, if fomething be fet againft the door, juft fufficient, when the plate is in, to keep the door nearly fhut, by refifting the preffure of the air that would force it open : Then, when the plate is drawn out, the door will be forced open by the increafed preffure of the outward cold air endeavouring to get in to fupply the place of the warm air, that now paffes out of the room to go up the chimney. In our common open chimneys, half the fuel is wafted, and its effect loft, the air it has warmed being immediately drawn off. Several of my acquaintance having feen this fimple machine in my room, have imitated

it

it at their own houses, and it seems likely to become pretty common. I describe it thus particularly to you, because I think it would be useful in *Boston*, where firing is often dear.

Mentioning chimneys puts me in mind of a property I formerly had occasion to observe in them, which I have not found taken notice of by others; it is, that in the summer time, when no fire is made in the chimneys, there is, nevertheless, a regular draft of air through them; continually passing upwards, from about five or six o'clock in the afternoon, till eight or nine o'clock the next morning, when the current begins to slacken and hesitate a little, for about half an hour, and then sets as strongly down again, which it continues to do till towards five in the afternoon, then slackens and hesitates as before, going sometimes a little up, then a little down, till in about half an hour it gets into a steady upward current for the night, which continues till eight or nine the next day; the hours varying a little as the days lengthen and shorten, and sometimes varying from sudden changes in the weather; as if, after being long warm, it should begin to grow cool about noon, while the air was coming down the chimney, the current will then change earlier than the usual hour, &c.

This property in chimneys I imagine we might turn to some account, and render improper, for the future, the old saying, *as useless as a chimney in summer.* If the opening of the chimney, from the breast down to the hearth, be closed by a slight moveable frame, or two in the manner

C c c   of

of doors, covered with canvas, that will let the air through, but keep out the flies; and another little frame set within upon the hearth, with hooks on which to hang joints of meat, fowls, &c. wrapt well in wet linen cloths, three or four fold, I am confident that if the linen is kept wet, by sprinkling it once a day, the meat would be so cooled by the evaporation, carried on continually by means of the passing air, that it would keep a week or more in the hottest weather. Butter and milk might likewise be kept cool, in vessels or bottles covered with wet cloths. A shallow tray, or keeler, should be under the frame to receive any water that might drip from the wetted cloths. I think, too, that this property of chimneys might, by means of smoak-jack vanes, be applied to some mechanical purposes, where a small but pretty constant power only is wanted.

If you would have my opinion of the cause of this changing current of air in chimneys, it is, in short, as follows. In summer time there is generally a great difference in the warmth of the air at mid-day and midnight, and, of course, a difference of specific gravity in the air, as the more it is warmed the more it is rarefied. The funnel of a chimney being for the most part surrounded by the house, is protected, in a great measure, from the direct action of the sun's rays, and also from the coldness of the night air. It thence preserves a middle temperature between the heat of the day, and the coldness of the night. This middle temperature it communicates to the air contained in it. If

the

the ftate of the outward air be cooler than that in the fun-nel of the chimney, it will, by being heavier, force it to rife, and go out at the top. What fupplies its place from below, being warmed, in its turn, by the warmer funnel, is likewife forced up by the colder and weightier air below, and fo the current is continued till the next day, when the fun gradually changes the ftate of the outward air, makes it firft as warm as the funnel of the chimney can make it, (when the current begins to hefitate) and afterwards warm-er. Then the funnel being cooler than the air that comes into it, cools that air, makes it heavier than the outward air ; of courfe it defcends ; and what fucceeds it from a-bove, being cool'd in its turn, the defcending current con-tinues till towards evening, when it again hefitates and changes its courfe, from the change of warmth in the outward air, and the nearly remaining fame middle tem-perature in the funnel.

Upon this principle, if a houfe were built behind *Beacon-hill*, an adit carried from one of the doors into the hill ho-rizontally, till it met with a perpendicular fhaft funk from its top, it feems probable to me, that thofe who lived in the houfe, would conftantly, in the heat even of the calmeft day, have as much cool air paffing through the houfe, as they fhould chufe ; and the fame, though reverfed in its current, during the ftilleft night.

I think, too, this property might be made of ufe to mi-ners ; as where feveral fhafts or pits are funk perpendicu-

larly

larly into the earth, communicating at bottom by horizontal paſſages, which is a common caſe, if a chimney of thirty or forty feet high were built over one of the ſhafts, or ſo near the ſhaft, that the chimney might communicate with the top of the ſhaft, all air being excluded but what ſhould paſs up or down by the ſhaft, a, conſtant change of air would, by this means, be produced in the paſſages below, tending to ſecure the workmen from thoſe damps which ſo frequently incommode them. For the freſh air would be almoſt always going down the open ſhaft, to go up the chimney, or down the chimney to go up the ſhaft. Let me add one obſervation more, which is, That if that part of the funnel of a chimney, which appears above the roof of a houſe, be pretty long, and have three of its ſides expoſed to the heat of the ſun ſucceſſively, *viz.* when he is in the eaſt, in the ſouth, and in the weſt ; while the north ſide is ſheltered by the building from the cool northerly winds. Such a chimney will often be ſo heated by the ſun, as to continue the draft ſtrongly upwards, through the whole twenty-four hours, and often for many days together. If the outſide of ſuch a chimney be painted black, the effect will be ſtill greater, and the current ſtronger.

I am, dear Sir, yours, &c. B. F.

LET-

LETTER XXXIV.

To Dr. *H.* at *London.*

S I R, *Craven-street, June* 7, 1759.

I NOW return the smallest of your two *Tourmalins,* with hearty thanks for your kind present of the other, which, though I value highly for its rare and wonderful properties, I shall ever esteem it more for the friendship I am honoured with by the giver.

I hear that the negative electricity of one side of the *Tourmalin,* when heated, is absolutely denied, (and all that has been related of it, ascribed to prejudice in favour of a system) by some ingenious gentlemen abroad, who profess to have made the experiments on the stone with care and exactness. The experiments have succeeded differently with me ; yet I would not call the accuracy of those gentlemen in question. Possibly the Tourmalins they have tried were not properly cut ; so that the positive and negative powers were obliquely placed, or in some manner whereby their effects were confused, or the negative part more easily supplied by the positive. Perhaps the lapidaries who have hitherto cut these stones, had no regard to the situation of the two powers, but chose to make the faces of the stone where they could obtain the greatest breadth, or some other advantage in the form. If any of

thefe

thefe ftones, in their natural ftate, can be procured here, I think it would be right to endeavour finding, before they are cut, the two fides that contain the oppofite powers, and make the faces there. Poffibly, in that cafe, the effects might be ftronger, and more diftinct ; for though both thefe ftones that I have examined have evidently the two properties, yet, without the full heat given by boiling water, they are fomewhat confufed ; the virtue feems ftrongeft towards one end of the face; and in the middle, or near the other end, fcarce difcernible ; and the negative, I think, always weaker than the pofitive.

I have had the large one new cut, fo as to make both fides alike, and find the change of form has made no change of power, but the properties of each fide remain the fame as I found them before. It is now fet in a ring in fuch a manner as to turn on an axis, that I may conveniently, in making experiments, come at both fides of the ftone. The little rim of gold it is fet in, has made no alteration in its effects. The warmth of my finger, when I wear it, is fufficient to give it fome degree of electricity, fo that it is always ready to attract light bodies.

The following experiments have fatisfied me that M. Ӕpinus's account of the pofitive and negative ftates of the oppofite fides of the heated Tourmalin, is well founded.

I heated the large ftone in boiling water.

As foon as it was dry, I brought it near a very fmall cork ball, that was fufpended by a filk thread.

The

The ball was attracted by one face of the ftone, which I call A, and then repelled.

The ball in that ftate was alfo repelled by the pofitively charg'd wire of a phial, and attracted by the other fide of the ftone, B.

The ftone being a-frefh heated, and the fide B brought near the ball, it was firft attracted, and prefently after re-pelled by that fide.

In this fecond ftate it was repelled by the negatively charged wire of a phial.

Therefore, if the principles now generally received, rela-ting to pofitive and negative electricity, are true, the fide A of the large ftone, when the ftone is heated in water, is in a pofitive ftate of electricity ; and the fide B, in a negative ftate.

The fame experiments being made with the fmall ftone ftuck by one edge on the end of a fmall glafs tube, with fealing-wax, the fame effects are produced. The flat fide of the fmall ftone gives the figns of pofitive electricity ; the high fide gives the figns of negative electricity.

Again ;

I fufpended the fmall ftone by a filk thread.

I heated it as it hung, in boiling water.

I heated the large one in boiling water.

Then I brought the large ftone near to the fufpended fmall one.

Which

Which immediately turned its flat fide to the fide B of the large ftone, and would cling to it.

I turned the ring, fo as to prefent the fide A of the large ftone, to the flat fide of the fmall one.

The flat fide was repelled, and the fmall ftone, turning quick, applied its high fide to the fide A of the large one.

This was precifely what ought to happen, on the fuppofition that the flat fide of the fmall ftone, when heated in water, is pofitive, and the high fide negative ; the fide A of the large ftone pofitive, and the fide B negative.

The effect was apparently the fame as would have been produced, if one magnet had been fufpended by a thread, and the different poles of another brought alternately near it.

I find that the face A, of the large ftone, being coated with leaf-gold, (attach'd by the white of an egg, which will bear dipping in hot water) becomes quicker and ftronger in its effect on the cork-ball, repelling it the inftant it comes in contact ; which I fuppofe to be occafioned by the united force of different parts of the face, collected and acting together through the metal.

I am, &c.   B. F.

LET-

LETTER XXXV.

To Mr. P. F. *in* Newport.

S I R, *London, May* 7, 1760.

* * * * * * It has, indeed, as you obferve, been the opinion of fome very great naturalifts, that the fea is falt only from the diffolution of mineral or rock falt, which its waters happened to meet with. But this opinion takes it for granted that all water was originally frefh, of which we can have no proof. I own I am inclined to a different opinion, and rather think all the water on this globe was originally falt, and that the frefh water we find in fprings and rivers, is the produce of diftillation. The fun raifes the vapours from the fea, which form clouds, and fall in rain upon the land, and fprings and rivers are form-ed of that rain.——As to the rock-falt found in mines, I conceive, that inftead of communicating its faltnefs to the fea, it is itfelf drawn from the fea, and that of courfe the fea is now frefher than it was originally. This is only a-nother effect of nature's diftillery, and might be performed various ways.

It is evident from the quantities of fea-fhells, and the bones and teeth of fifhes found in high lands, that the fea has formerly covered them. Then, either the fea has been higher than it now is, and has fallen away from thofe high

D d d lands,

lands ; or they have been lower than they are, and were lifted up out of the water to their prefent height, by fome internal mighty force, fuch as we ftill feel fome remains of, when whole continents are moved by earthquakes. In either cafe it may be fuppofed that large hollows, or valleys among hills, might be left filled with fea-water, which evaporating, and the fluid part drying away in a courfe of years, would leave the falt covering the bottom ; and that falt coming afterwards to be covered with earth, from the neighbouring hills, could only be found by digging through that earth. Or, as we know from their effects, that there are deep fiery caverns under the earth, and even under the fea, if at any time the fea leaks into any of them, the fluid parts of the water muft evaporate from that heat, and pafs off through fome vulcano, while the falt remains, and by degrees, and continual accretion, becomes a great mafs. Thus the cavern may at length be filled, and the volcano connected with it ceafe burning, as many it is faid have done ; and future miners penetrating fuch cavern, find what we call a falt mine.——This is a fancy I had on vifiting the falt-mines at *Northwich*, with my fon. I fend you a piece of the rock-falt which he brought up with him out of the mine. * * * * * *

I am, Sir, &c.   B. F.

LET-

L E T T E R XXXVI.

To Mr. ALEXANDER SMALL, *London.*

Dear Sir,   *May* 12, 1760.

Agreeable to your requeft, I fend you my reafons for thinking that our North-Eaft ftorms in *North-America* begin firft, in point of time, in the South-Weft parts : That is to fay, the air in *Georgia*, the fartheft of our colonies to the South-Weft, begins to move South-Wefterly befere the air of *Carolina*, which is the next colony North-Eaftward ; the air of *Carolina* has the fame motion before the air of *Virginia*, which lies ftill more North-Eaftward ; and fo on North-Eafterly through *Penfylvania, New-York, New-England, &c.* quite to *Newfoundland.*

Thefe North-Eaft ftorms are generally very violent, continue fometimes two or three days, and often do confiderable damage in the harbours along the coaft. They are attended with thick clouds and rain.

What firft gave me this idea, was the following circumftance. About twenty years ago, a few more or lefs, I cannot from my memory be certain, we were to have an eclipfe of the moon at *Philadelphia*, on a *Friday* evening, about nine o'clock. I intended to obferve it, but was prevented by a North-Eaft ftorm, which came on about fe-

ven,

ven, with thick clouds as ufual, that quite obfcured the
whole hemifphere. Yet when the poft brought us the
Bofton news-paper, giving an account of the effects of the
fame ftorm in thofe parts, I found the beginning of the
eclipfe had been well obferved there, though *Bofton* lies
N. E. of *Philadelphia* about 400 miles. This puzzled
me, becaufe the ftorm began with us fo foon as to prevent
any obfervation, and being a N. E. ftorm, I imagined it
muft have began rather fooner in places farther to the North
Eaftward, than it did at *Philadelphia*. I therefore mentioned it in a letter to my brother who lived at *Bofton* ; and
he informed me the ftorm did not begin with them till
near eleven o'clock, fo that they had a good obfervation of
the eclipfe : And upon comparing all the other accounts I
received from the feveral colonies, of the time of beginning of the fame ftorm, and fince that of other ftorms of
the fame kind, I found the beginning to be always later
the farther North-Eaftward. I have not my notes with me
here in *England,* and cannot, from memory, fay the proportion of time to diftance, but I think it is about an hour
to every hundred miles.

From thence I formed an idea of the caufe of thefe
ftorms, which I would explain by a familiar inftance or
two.——Suppofe a long canal of water ftopp'd at the end
by a gate. The water is quite at reft till the gate is open,
then it begins to move out through the gate; the water
next the gate is firft in motion, and moves towards the

gate ;

gate ; the water next to that firft water moves next, and fo on fucceffively, till the water at the head of the canal is in motion, which is laft of all. In this cafe all the water moves indeed towards the gate, but the fucceffive times of beginning motion are the contrary way, *viz.* from the gate backwards to the head of the canal. Again, fuppofe the air in a chamber at reft, no current through the room till you make a fire in the chimney. Immediately the air in the chimney being rarefied by the fire, rifes ; the air next the chimney flows in to fupply its place, moving towards the chimney ; and, in confequence, the reft of the air fucceffively, quite back to the door. Thus to produce our North Eaft ftorms, I fuppofe fome great heat and rarefaction of the air in or about the Gulph of *Mexico* ; the air thence rifing has its place fupplied by the next more northern, cooler, and therefore denfer and heavier, air ; that, being in motion, is followed by the next more northern air, *&c. &c.* in a fucceffive current, to which current our coaft and inland ridge of mountains give the direction of North-Eaft, as they lie N. E. and S. W.

This I offer only as an hypothefis to account for this particular fact ; and, perhaps, on farther examination, a better and truer may be found. I do not fuppofe all ftorms generated in the fame manner. Our North-Weft thundergufts in *America* I know are not ; but of them I have written my opinion fully in a paper which you have feen.

I am, &c.   B. F.

L E T-

LETTER XXXVII.

From Mr. KINNERSLEY.

S I R,   *Philadelphia, March* 12, 1761.

HAVING lately made the following experiments, I
very chearfully communicate them, in hopes of
giving you fome degree of pleafure, and exciting you to
further explore your favourite, but not quite exhaufted fubject, ELECTRICITY.

I placed myfelf on an electric ftand, and, being well e-
lectrifed, threw my hat to an unelectrifed perfon, at a confiderable diftance, on another ftand, and found that the hat
carried fome of the electricity with it; for, upon going
immediately to the perfon who received it, and holding a
flaxen thread near him, I perceived he was electrifed fufficiently to attract the thread.

I then fufpended, by filk, a broad plate of metal, and e-
lectrifed fome boiling water under it, at about four feet diftance, expecting that the vapour, which afcended plentifully
to the plate, would, upon the principle of the foregoing
experiment, carry up fome of the electricity with it ; but
was at length fully convinced, by feveral repeated trials,
that it left all its fhare thereof behind. This I know not
how to account for ; but does it not feem to corroborate

your

your hypothefis, That the vapours of which the clouds are formed, leave their fhare of electricity behind, in the common ftock, and afcend in the negative ftate?

I put boiling water into a coated *Florence* flafk, and found that the heat fo enlarged the pores of the glafs, that it could not be charged. The electricity paffed through as readily, to all appearance, as through metal; the charge of a three pint bottle went freely through, without injuring the flafk in the leaft. When it became almoft cold, I could charge it as ufual. Would not this experiment convince the Abbe *Nollet* of his egregious miftake? For while the electricity went fairly through the glafs, as he contends it always does, the glafs could not be charged at all.

I took a flender piece of cedar, about eighteen inches long, fixed a brafs cap in the middle, thruft a pin horizontally and at right angles, through each end, (the points in contrary directions) and hung it, nicely ballanc'd, like the needle of a compafs, on a pin, about fix inches long, fixed in the center of an electric ftand. Then, electrifing the ftand, I had the pleafure of feeing what I expected; the wooden needle turned round, carrying the pins with their heads foremoft. I then electrifed the ftand negatively, expecting the needle to turn the contrary way, but was extremely difappointed, for it went ftill the fame way as before. When the ftand was electrifed pofitively, I fuppofe that the natural quantity of electricity in the air being increafed on one fide, by what iffued from the points, the

needle

needle was attracted by the leſſer quantity on the other ſide. When electriſed negatively, I ſuppoſe that the natural quantity of electricity in the air was diminiſhed near the points; in conſequence whereof, the equilibrium being deſtroyed, the needle was attracted by the greater quantity on the oppoſite ſide.

The doctrine of repulſion, in electriſed bodies, I begin to be ſomewhat doubtful of. I think all the phænomena on which it is founded, may be well enough accounted for without it. Will not cork balls, electriſed negatively, ſeparate as far as when electriſed poſitively? And may not their ſeparation in both caſes be accounted for upon the ſame principle, namely, the mutual attraction of the natural quantity in the air, and that which is denſer or rarer in the cork balls? it being one of the eſtabliſhed laws of this fluid, that quantities of different denſities ſhall mutually attract each other, in order to reſtore the equilibrium.

I can ſee no reaſon to conclude that the air has not its ſhare of the common ſtock of electricity, as well as glaſs, and, perhaps, all other electrics *per ſe*. For though the air will admit bodies to be electriſed in it either poſitively or negatively, and will not readily carry off the redundancy in the one caſe, or ſupply the deficiency in the other, yet let a perſon in the negative ſtate, out of doors in the dark, when the air is dry, hold, with his arm extended, a long ſharp needle, pointing upwards, and he will ſoon be convinced that electricity may be drawn out of the air; not

very

very plentifully, for, being a bad conductor, it feems loth to part with it, but yet fome will evidently be collected. The air near the perfon's body having lefs than its natural quantity, will have none to fpare; but, his arm being extended, as above, fome will be collected from the remoter air, and will appear luminous, as it converges to the point of the needle.

Let a perfon electrifed negatively prefent the point of a needle, horizontally, to a cork ball, fufpended by filk, and the ball will be attracted towards the point, till it has parted with fo much of its natural quantity of electricity as to be in the negative ftate in the fame degree with the perfon who holds the needle; then it will recede from the point, being, as I fuppofe, attracted the contrary way by the electricity of greater denfity in the air behind it. But, as this opinion feems to deviate from electrical orthodoxy, I fhould be glad to fee thefe phænomena better accounted for by your fuperior and more penetrating genius.

Whether the electricity in the air, in clear dry weather, be of the fame denfity at the height of two or three hundred yards, as near the furface of the earth, may be fatisfactorily determined by your old experiment of the kite. The twine fhould have, throughout, a very fmall wire in it, and the ends of the wire, where the feveral lengths are united, ought to be tied down with a waxed thread, to prevent their acting in the manner of points. I have tried the experiment twice, when the air was as dry as we ever have

it,

it, and fo clear that not a cloud could be feen, and found the twine each time in a fmall degree electrifed pofitively. The kite had three metalline points fixed to it; one on the top, and one on each fide. That the twine was electrifed, appeared by the feparating of two fmall cork balls, fufpended on the twine by fine flaxen threads, juft above where the filk was tied to it, and fheltered from the wind. That the twine was electrifed pofitively, was proved, by applying to it the wire of a charged bottle, which caufed the balls to feparate further, without firft coming nearer together. This experiment fhewed that the electricity in the air, at thofe times, was denfer above than below. But that cannot be always the cafe; for you know we have frequently found the thunder clouds in the negative ftate, attracting electricity from the earth; which ftate, it is probable, they are always in when firft formed, and till they have received a fufficient fupply. How they come afterwards, towards the latter end of the guft, to be in the pofitive ftate, which is fometimes the cafe, is a fubject for further enquiry.

After the above experiments with the wooden needle, I formed a crofs, of two pieces of wood, of equal length, interfecting each other at right angles in the middle, hung it horizontally upon a central pin, and fet a light horfe, with his rider, upon each extremity; whereupon, the whole being nicely balanced, and each courfer urged on by

an

PLATE VI.

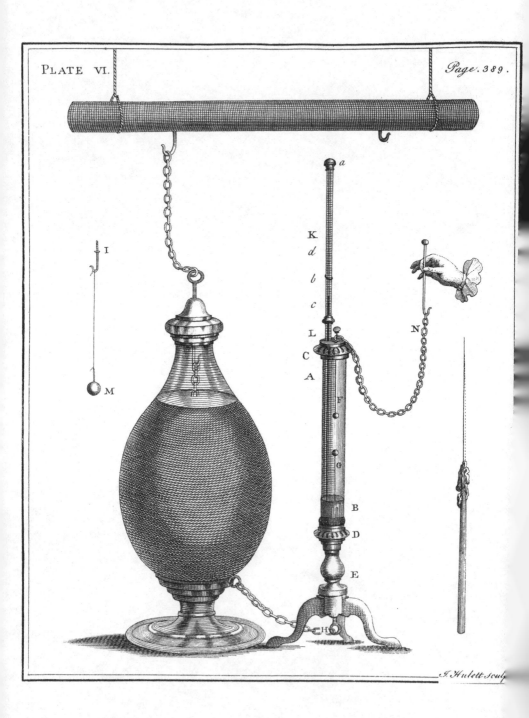

J. Hulett Sculp.

an electrified point inftead of a pair of fpurs, I was entertained with an electrical horfe-race.

I have contrived an electrical air thermometer, and made feveral experiments with it, that have afforded me much fatisfaction and pleafure. It is extremely fenfible of any alteration in the ftate of the included air, and fully determines that controverted point, Whether there be any heat in the electric fire? By the enclofed draught, and the following defcription, you will readily apprehend the conftruction of it.

A B is a glafs tube, about eleven inches long, and one inch diameter in the bore. It has a brafs feril, cemented on each end, with a top and bottom part, C and D, to be fcrewed on, air-tight, and taken off at pleafure. In the center of the bottom part D, is a male fcrew, which goes into a brafs nut, in the mahogany pedeftal E. The wires F and G, are for the electric fire to pafs through, darting from one to the other. The wire G extends through the pedeftal to H, and may be raifed and lowered by means of a male fcrew on it. The wire F may be taken out, and the hook I be fcrewed into its place. K is a glafs tube, with a fmall bore, open at both ends, cemented in the brafs tube L, which fcrews into the top part C. The lower end of the tube K is immerfed in water, coloured with cochineal, at the bottom of the tube A B. (I ufed, at firft, coloured fpirits of wine, but in one experiment I made, it took fire.) On the top of the tube K is cemented, for or-

nament,

nament, a brafs feril, with a head fcrewed on it, which has a fmall air-hole through its fide, at *a*. The wire *b*, is a fmall round fpring, that embraces the tube K, fo as to ftay wherever it is placed. The weight M is to keep ftrait whatever may be fufpended in the tube A B, on the hook I. Air muft be blown through the tube K, into the tube A B, till enough is intruded to raife, by its elaftic force, a column of the coloured water in the tube K, up to *c*, or thereabouts ; and then, the gage-wire *b*, being flipt down to the top of the column, the thermometer is ready for ufe.

I fet the thermometer on an electric ftand, with the chain N fixed to the prime conductor, and kept it well e-lectrifed a confiderable time ; but this produced no fen-fible effect ; which fhews, that the electric fire, when in a ftate of reft, has no more heat than the air, and other matter wherein it refides.

When the wires F and G are in contact, a large charge of electricity fent through them, even that of my cafe of five and thirty bottles, containing above thirty fquare feet of coated glafs, will produce no rarefaction of the air in-cluded in the tube A B; which fhews that the wires are not heated by the fire's paffing through them.

When the wires are about two inches apart, the charge of a three pint bottle, darting from one to the other, ra-refies the air very evidently ; which fhews, I think, that

the

the electric fire muft produce heat in itfelf, as well as in the air, by its rapid motion.

The charge of one of my glafs jars, (which will contain about five gallons and a half, wine meafure) darting from wire to wire, will, by the difturbance it gives the air, repelling it in all directions, raife the column in the tube K, up to *d*, or thereabouts ; and the charge of the abovementioned cafe of bottles will raife it to the top of the tube. Upon the air's coalefcing, the column, by its gravity, inftantly fubfides, till it is in equilibrio with the rarefied air ; it then gradually defcends, as the air cools, and fettles where it ftood before. By carefully obferving at what height above the gage-wire *b*, the defcending column firft ftops, the degree of rarefaction is difcovered, which, in great explofions, is very confiderable.

I hung in the thermometer, fucceffively, a ftrip of wet writing paper, a wet flaxen and woolen thread, a blade of green grafs, a filament of green wood, a fine filver thread, a very fmall brafs wire, and a ftrip of gilt paper ; and found that the charge of the above-mentioned glafs jar, paffing through each of thefe, efpecially the laft, produced heat enough to rarefy the air very perceptibly.

I then fufpended, out of the thermometer, a piece of fmall harpfichord wire, about twenty-four inches long, with a pound weight at the lower end, and fent the charge of the cafe of five and thirty bottles through it, whereby I difcovered a new method of wire-drawing. The wire

was

was red hot the whole length, well annealed, and above an inch longer than before. A fecond charge melted it; it parted near the middle, and meafured, when the ends were put together, four inches longer than at firft. This experiment, I remember, you propofed to me before you left *Philadelphia*; but I never tried it till now. That I might have no doubt of the wire's being *hot* as well as red, I repeated the experiment on another piece of the fame wire, encompaffed with a goofe-quill, filled with loofe grains of gun-powder; which took fire as readily as if it had been touched with a red hot poker. Alfo tinder, tied to another piece of the wire, kindled by it. I tried a wire about three times as big, but could produce no fuch effects with that.

Hence it appears that the electric fire, though it has no fenfible heat when in a ftate of reft, will, by its violent motion, and the refiftance it meets with, produce heat in other bodies when paffing through them, provided they be fmall enough. A large quantity will pafs through a large wire, without producing any fenfible heat; when the fame quantity paffing through a very fmall one, being there confined to a narrower paffage, the particles crowding clofer together, and meeting with greater refiftance, will make it red hot, and even melt it.

Hence lightning does not melt metal by a cold fufion, as we formerly fuppofed; but, when it paffes through the blade of a fword, if the quantity be not very great, it may

heat

heat the point fo as to melt it, while the broadeft and thick-eft part may not be fenfibly warmer than before.

And when trees or houfes are fet on fire by the dread-ful quantity which a cloud, or the earth, fometimes difchar-ges, muft not the heat, by which the wood is firft kindled, be generated by the lightning's violent motion, through the refifting combuftible matter ?

If lightning, by its rapid motion, produces heat in *itfelf*, as well as in other bodies, (and that it does I think is evi-dent from fome of the foregoing experiments made with the thermometer) then its fometimes fingeing the hair of animals killed by it, may eafily be accounted for. And the reafon of its not always doing fo, may, perhaps, be this; The quantity, though fufficient to kill a large animal, may fometimes not be great enough, or not have met with re-fiftance enough, to become, by its motion, burning hot.

We find that dwelling-houfes, ftruck with lightning, are feldom fet on fire by it; but when it paffes through barns, with hay or ftraw in them, or ftore-houfes, contain-ing large quantities of hemp, or fuch like matter, they fel-dom, if ever, efcape a conflagration; which may, perhaps, be owing to fuch combuftibles being apt to kindle with a lefs degree of heat than is neceffary to kindle wood.

We had four houfes in this city, and a veffel at one of the wharfs, ftruck and damaged by lightning laft fummer. One of the houfes was ftruck twice in the fame ftorm. But I have the pleafure to inform you, that your method

of

of preventing fuch terrible difafters, has, by a fact which had like to have efcaped our knowledge, given a very convincing proof of its great utility, and is now in higher repute with us than ever.

Hearing, a few days ago, that Mr. *William Weft*, merchant in this city, fufpected that the lightning in one of the thunder-ftorms laft fummer, had paffed through the iron conductor which he had provided for the fecurity of his houfe; I waited on him, to enquire what ground he might have for fuch fufpicion. Mr. *Weft* informed me, that his family and neighbours were all ftunned with a very terrible explofion, and that the flafh and crack were feen and heard at the fame inftant. Whence he concluded, that the lightning muft have been very near, and, as no houfe in the neighbourhood had fuffered by it, that it muft have paffed through his conductor. Mr. *White*, his clerk, told me that he was fitting, at the time, by a window, about two feet diftant from the conductor, leaning againft the brick wall with which it was in contact; and that he felt a fmart fenfation, like an electrick fhock, in that part of his body which touched the wall. Mr. *Weft* further informed me, that a perfon of undoubted veracity affured him, that, being in the door of an oppofite houfe, on the other fide of *Water-ftreet*, (which you know is but narrow) he faw the lightning diffufed over the pavement, which was then very wet with rain, to the diftance of two or three yards from the foot of the conductor; and that

another

another perfon of very good credit told him, that he being a few doors off on the other fide of the ftreet, faw the lightning above, darting in fuch direction that it appeared to him to be directly over that pointed rod.

Upon receiving this information, and being defirous of further fatisfaction, there being no traces of the lightning to be difcovered in the conductor, as far as we could examine it below, I propofed to Mr. *Weft* our going to the top of the houfe, to examine the pointed rod, affuring him, that if the lightning had paffed through it, the point muft have been melted ; and, to our great fatisfaction, we found it fo. This iron rod extended in height about nine feet and a half above a ftack of chimneys to which it was fixed, (though I fuppofe three or four feet would have been fufficient.) It was fomewhat more than half an inch diameter in the thickeft part, and tapering to the upper end. The conductor, from the lower end of it to the earth, confifted of fquare iron nail-rods, not much above a quarter of an inch thick, connected together by interlinking joints. It extended down the cedar roof to the eaves, and from thence down the wall of the houfe, four ftory and a half, to the pavement in *Water-ftreet*, being faftened to the wall, in feveral places, by fmall iron hooks. The lower end was fixed to a ring, in the top of an iron ftake that was drove about four or five feet into the ground.

The above-mentioned iron rod had a hole in the top of it, about two inches deep, wherein was inferted a brafs

F f f wire,

wire, about two lines thick, and, when firſt put there, a-
bout ten inches long, terminating in a very acute point;
but now its whole length was no more than ſeven inches
and a half, and the top very blunt. Some of the metal
appears to be miſſing, the ſlendereſt part of the wire being, as I ſuſpect, conſumed into ſmoke. But ſome of it,
where the wire was a little thicker, being only melted by
the lightning, ſunk down, while in a fluid ſtate, and formed a rough irregular cap, lower on one ſide than the other,
round the upper end of what remained, and became intimately united therewith.

This was all the damage that Mr. *Weſt* ſuſtained by a
terrible ſtroke of lightning;—a moſt convincing proof of
the great utility of this method of preventing its dreadful
effects. Surely it will now be thought as expedient to provide conductors for the lightning, as for the rain.

Mr. *Weſt* was ſo good as to make me a preſent of the
melted wire, which I keep as a great curioſity, and long
for the pleaſure of ſhewing it to you. In the mean time,
I beg your acceptance of the beſt repreſentation I can give
of it, which you will find by the ſide of the thermometer,
drawn in its full dimenſions as it now appears. The dotted
lines above are intended to ſhew the form of the wire before the lightning melted it.

And now, Sir, I moſt heartily congratulate you on the
pleaſure you muſt have in finding your great and well-
grounded expectations ſo far fulfilled. May this method
of

of fecurity from the deftructive violence of one of the moft awful powers of nature, meet with fuch further fuccefs, as to induce every good and grateful heart to blefs God for the important difcovery! May the benefit thereof be diffufed over the whole globe! May it extend to the lateft pofterity of mankind, and make the name of *F R A N K-L I N*, like that of *N E W T O N*, *immortal*.

I am, Sir, with fincere refpect,

Your moft obedient and moft humble fervant,

EBEN. KINNERSLEY.

L E T T E R XXXVIII.

To Mr. KINNERSLEY, in anfwer to the foregoing.

S I R,   *London, Feb.* 20, 1762.

I Received your ingenious letter of the 12th of *March* laft, and thank you cordially for the account you give me of the new experiments you have lately made in Electricity.—It is a fubject that ftill affords me pleafure, though of late I have not much attended to it.

Your fecond experiment, in which you attempted, without fuccefs, to communicate pofitive electricity by vapour

F f f 2   afcending

afcending from electrified water, reminds me of one I formerly made, to try if negative electricity might be produced by evaporation only. I placed a large heated brafs
plate, containing four or five fquare feet, on an electric
ftand ; a rod of metal, about four feet long, with a bullet
at its end, extended from the plate horizontally. A light
lock of cotton, fufpended by a fine thread from the cieling, hung oppofite to, and within an inch of the bullet. I
then fprinkled the heated plate with water, which arofe
faft from it in vapour. If vapour fhould be difpofed to
carry off the electrical, as it does the common fire from
bodies, I expected the plate would, by lofing fome of its
natural quantity, become negatively electrifed. But I
could not perceive, by any motion in the cotton, that it
was at all affected ; nor by any feparation of fmall cork-
balls fufpended from the plate, could it be obferved that
the plate was in any manner electrified. Mr. *Canton* here
has alfo found, that two tea-cups, fet on electric ftands,
and filled, one with boiling, the other with cold water,
and equally electrified, continued equally fo, notwithftanding the plentiful evaporation from the hot water. Your
experiment and his agreeing, fhow another remarkable
difference between electric and common fire. For the
latter quits moft readily the body that contains it, where
water, or any other fluid, is evaporating from the furface
of that body, and efcapes with the vapour. Hence the
method long in ufe in the eaft, of cooling liquors, by

wrapping

wrapping the bottles round with a wet cloth, and expof-
ing them to the wind. Dr. *Cullen,* of *Edinburgh,* has
given fome experiments of cooling by evaporation ; and
I was prefent at one made by Dr. *Hadley,* then profeffor
of chemiftry at *Cambridge,* when, by repeatedly wetting
the ball of a thermometer with fpirit, and quickening the
evaporation by the blaft of a bellows, the mercury fell
from 65, the ftate of warmth in the common air, to 7,
which is 22 degrees below freezing ; and, accordingly,
from fome water mixed with the fpirit, or from the breath
of the affiftants, or both, ice gathered in fmall fpicula
round the ball, to the thicknefs of near a quarter of an inch.
To fuch a degree did the mercury lofe the fire it before
contained, which, as I imagine, took the opportunity of
efcaping, in company with the evaporating particles of the
fpirit, by adhering to thofe particles.

Your experiment of the *Florence* flafk, and boiling water, is very curious. I have repeated it, and found it to fucceed as you defcribe it, in two flafks out of three. The
third would not charge when filled with either hot or cold
water. I repeated it, becaufe I remembered I had once
attempted to make an electric bottle of a *Florence* flafk,
filled with cold water, but could not charge it at all ;
which I then imputed to fome imperceptible cracks in
the fmall, extremely thin bubbles, of which that glafs is
full, and I concluded none of that kind would do. But
you have fhewn me my miftake.—Mr. *Wilfon* had formerly

merly acquainted us, that red hot glaſs would conduct e-
lectricity ; but that ſo ſmall a degree of heat as that communicated by boiling water, would ſo open the pores of
extremely thin glaſs, as to ſuffer the electric fluid freely to
paſs, was not before known. Some experiments ſimilar
to yours, have, however, been made here, before the receipt of your letter, of which I ſhall now give you an account.

I formerly had an opinion that a *Leyden* bottle, charg'd
and then ſeal'd hermetically, might retain its electricity
for ever ; but having afterwards ſome ſuſpicion that poſſibly that ſubtil fluid might, by ſlow imperceptible degrees, ſoak through the glaſs, and in time eſcape, I re-
queſted ſome of my friends, who had conveniences for
doing it, to make trial, whether, after ſome months, the
charge of a bottle ſo ſealed would be ſenſibly diminiſhed.
Being at *Birmingham*, in *September* 1760, Mr. *Bolton* of
that place opened a bottle that had been charged, and its
long tube neck hermetically ſealed in the *January* preceding. On breaking off the end of the neck, and introducing a wire into it, we found it poſſeſſed of a conſiderable
quantity of electricity, which was diſcharged by a ſnap
and ſpark. This bottle had lain near ſeven months on a
ſhelf, in a cloſet, in contact with bodies that would undoubtedly have carried off all its electricity, if it could have
come readily through the glaſs. Yet as the quantity ma-
nifeſted by the diſcharge was not apparently ſo great as
might

might have been expected from a bottle of that fize well, charged, fome doubt remained whether part had efcaped while the neck was fealing, or had fince, by degrees, foaked through the glafs. But an experiment of Mr. *Canton's*, in which fuch a bottle was kept under water a week, without having its electricity in the leaft impaired, feems to fhow, that when the glafs is cold, though extremely thin, the electric fluid is well retained by it. As that ingenious and accurate experimenter made a difcovery, like yours, of the effect of heat in rendering thin glafs permeable by that fluid, it is but doing him juftice to give you his account of it, in his own words, extracted from his letter to me, in which he communicated it, dated *Oct.* 31, 1760, *viz.*

" Having procured fome thin glafs balls, of about an inch and a half in diameter, with ftems, or tubes, of eight or nine inches in length, I electrified them, fome pofitively on the infide, and others negatively, after the manner of charging the *Leyden* bottle, and fealed them hermetically. Soon after I applied the naked balls to my electrometer, and could not difcover the leaft fign of their being electrical ; but holding them before the fire, at the diftance of fix or eight inches, they became ftrongly electrical in a very fhort time, and more fo when they were cooling. Thefe balls will, every time they are heated, give the electrical fluid to, or take it from other bodies, according to the *plus* or *minus* ftate of it within them. Heating them frequently, I find will fenfibly diminifh their power ; but

keeping

keeping one of them under water a week, did not appear in the leaft degree to impair it. That which I kept under water, was charged on the 22d of *September* laft, was feveral times heated before it was kept in water, and has been heated frequently fince, and yet it ftill retains its virtue to a very confiderable degree. The breaking two of my balls accidentally, gave me an opportunity of meafuring their thicknefs, which I found to be between feven and eight parts in a thoufand of an inch.

A down feather, in a thin glafs ball, hermetically fealed, will not be affected by the application of an excited tube, or the wire of a charged vial, unlefs the ball be confiderably heated ; and if a glafs pane be heated till it begins to grow foft, and in that ftate be held between the wire of a charged vial, and the difcharging wire, the courfe of the electrical fluid will not be through the glafs, but on the furface, round by the edge of it."

By this laft experiment of Mr. *Canton*'s, it appears, that though by a moderate heat, thin glafs becomes, in fome degree, a conductor of electricity, yet, when of the thicknefs of a common pane, it is not, though in a ftate near melting, fo good a conductor as to pafs the fhock of a difcharged bottle. There are other conductors which fuffer the electric fluid to pafs through them gradually, and yet will not conduct a fhock. For inftance, a quire of paper will conduct through its whole length, fo as to electrify a perfon, who, ftanding on wax, prefents the paper to an electri-

fied

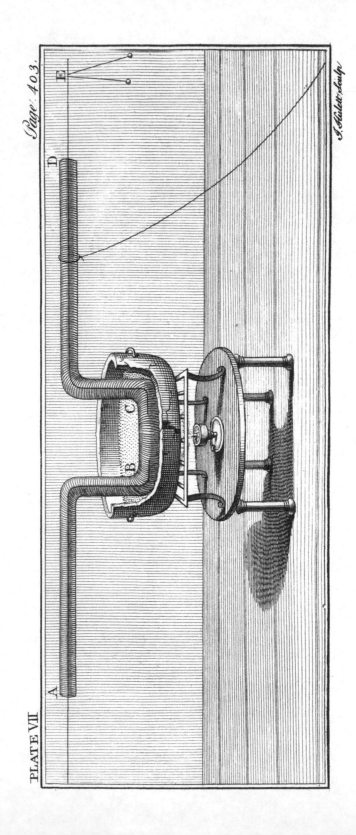

PLATE VII

A

B

D

E

J. Aikett sculp.

fied prime conductor; but it will not conduct a shock even through its thickness only; hence the shock either fails, or passes by rending a hole in the paper. Thus a sieve will pass water gradually, but a stream from a fire engine would either be stopped by it, or tear a hole through it.

It should seem, that to make glass permeable to the electric fluid. the heat should be proportioned to the thickness. You found the heat of boiling water, which is but 210, sufficient to render the extreme thin glass in a *Florence* flask permeable even to a shock.—Lord *Charles Cavendish*, by a very ingenious experiment, has found the heat of 400 requisite to render thicker glass permeable to the common current.

" A glass tube, (See *Plate* VI.) of which the part C B was solid, had wire thrust in each end, reaching to B and C.

A small wire was tied on at D, reaching to the floor, in order to carry off any electricity that might run along upon the tube.

The bent part was placed in an iron pot, filled with iron filings ; a thermometer was also put into the filings ; a lamp was placed under the pot ; and the whole was supported upon glass.

The wire A being electrified by a machine, before the heat was applied, the corks at E separated, at first upon the principle of the *Leyden* vial.

But after the part C B of the tube was heated to 600, the corks continued to separate, though you discharged the

electricity

electricity by touching the wire at E, the electrical machine continuing in motion.

Upon letting the whole cool, the effect remained till the thermometer was sunk to 400."

It were to be wished, that this noble philosopher would communicate more of his experiments to the world, as he makes many, and with great accuracy.

You know I have always look'd upon and mentioned the equal repulsion in cases of positive and of negative electricity, as a phænomenon difficult to be explained. I have sometimes, too, been inclined, with you, to resolve all into attraction; but besides that attraction seems in itself as unintelligible as repulsion, there are some appearances of repulsion that I cannot so easily explain by attraction; this for one instance. When the pair of cork balls are suspended by flaxen threads, from the end of the prime conductor, if you bring a rubbed glass tube near the conductor, but without touching it, you see the balls separate, as being electrified positively; and yet you have communicated no electricity to the conductor, for, if you had, it would have remained there, after withdrawing the tube; but the closing of the balls immediately thereupon, shews that the conductor has no more left in it than its natural quantity. Then again approaching the conductor with the rubbed tube, if, while the balls are separated, you touch with a finger that end of the conductor to which they hang, they will come together again. as being, with that part of the

conductor

conductor, brought to the fame ftate with your finger, *i. e.*
the natural ftate. But the other end of the conductor, near
which the tube is held, is not in that ftate, but in the negative ftate, as appears on removing the tube ; for then part
of the natural quantity left at the end near the balls, leaving that end to fupply what is wanting at the other, the
whole conductor is found to be equally in the negative ftate.
Does not this indicate that the electricity of the rubbed
tube had repelled the electric fluid, which was diffufed in
the conductor while in its natural ftate, and forced it to quit
the end to which the tube was brought near, accumulating
itfelf on the end to which the balls were fufpended ? I
own I find it difficult to account for its quitting that end,
on the approach of the rubbed tube, but on the fuppofition of repulfion ; for, while the conductor was in the fame
ftate with the air, *i. e.* the natural ftate, it does not feem to
me eafy to fuppofe, that an attraction fhould fuddenly
take place between the air and the natural quantity of the
electric fluid in the conductor, fo as to draw it to, and
accumulate it on the end oppofite to that approached by the
tube ; fince bodies, poffeffing only their natural quantity
of that fluid, are not ufually feen to attract each other, or to
affect mutually the quantities of electricity each contains.

There are likewife appearances of repulfion in other
parts of nature. Not to mention the violent force with
which the particles of water, heated to a certain degree,
feparate from each other, or thofe of gunpowder, when

touch'd

touch'd with the fmalleft fpark of fire, there is the feeming
repulfion between the fame poles of the magnet, a body
containing a fubtle moveable fluid, in many refpects analagous to the electric fluid.   If two magnets are fo fufpended by ftrings, as that their poles of the fame denomina-
tion are oppofite to each other, they will feparate, and continue fo ; or if you lay a magnetic fteel bar on a fmooth
table, and approach it with another parallel to it, the poles
of both in the fame pofition, the firft will recede from the
fecond fo as to avoid the contact,. and may thus be pufh'd
(or at leaft appear to be pufh'd) off the table.   Can this
be afcribed to the attraction of any furrounding body or
matter drawing them afunder, or drawing the one away
from the other ?   If not, and repulfion exifts in nature,
and in magnetifm, why may it not exift in electricity ?
We fhould not, indeed, multiply caufes in philofophy without neceffity ; and the greater fimplicity of your hypothefis would recommend it to me, if I could fee that all appearances might be folved by it.   But I find, or think I
find, the two caufes more convenient than one of them a-
lone.   Thus I would folve the circular motion of your
horizontal ftick, fupported on a pivot, with two pins at
their ends, pointing contrary ways, and moving in the fame
direction when electrified, whether pofitively or negatively: When pofitively, the air oppofite to the points being
electrifed pofitively, repels the points; when negatively,
the air oppofite the points being alfo, by their means, elec-
                                                        trifed

trified negatively, attraction takes place between the electricity in the air behind the heads of the pins, and the negative pins, and so they are, in this case, drawn in the same direction that in the other they were driven.—You see I am willing to meet you half way, a complaisance I have not met with in our brother *Nollet*, or any other hypothesis-maker, and therefore may value myself a little upon it, especially as they say I have some ability in defending even the wrong side of a question, when I think fit to take it in hand.

What you give as an established law of the electric fluid, " That quantities of different densities mutually attract " each other, in order to restore the equilibrium," is, I think, not well founded, or else not well express'd. Two large cork balls, suspended by silk strings, and both well and equally electrified, separate to a great distance. By bringing into contact with one of them, another ball of the same size, suspended likewise by silk, you will take from it half its electricity. It will then, indeed, hang at a less distance from the other, but the full and the half quantities will not appear to attract each other, that is, the balls will not come together. Indeed, I do not know any proof we have, that one quantity of electric fluid is attracted by another quantity of that fluid, whatever difference there may be in their densities. And, supposing in nature, a mutual attraction between two parcels of any kind of matter, it would be strange if this attraction should subsist strongly

while

while thofe parcels were unequal, and ceafe when more
matter of the fame kind was added to the fmalleft parcel,
fo as to make it equal to the biggeft.   By all the laws of
attraction in matter, that we are acquainted with, the attraction is ftronger in proportion to the increafe of the maffes, and never in proportion to the difference of the maffes.
I fhould rather think the law would be, " That the elec-
" tric fluid is attracted ftrongly by all other matter that we
" know of, while the parts of that fluid mutually rep⁻¹
" each other."   Hence its being equally diffufed (except in
particular circumftances)   throughout all other matter.
But this you jokingly call " electrical orthodoxy."   It is fo
with fome at prefent, but not with all; and, perhaps, it
may not always be orthodoxy with any body.   Opinions
are continually varying, where we cannot have mathematical evidence of the nature of things; and they muft vary.
Nor is that variation without its ufe, fince it occafions a
more thorough difcuffion, whereby error is often diffipated, true knowledge is encreafed, and its principles become
better underftood and more firmly eftablifhed.

Air fhould have, as you obferve,   its fhare of the
" common ftock of electricity, as well as glafs, and, per-
" haps, all other electrics *per fe*."   But I fuppofe, that,
like them, it does not eafily part with what it has, or receive more, unlefs when mix'd with fome non-electric, as
moifture for inftance, of which there is fome in our drieft
air.   This, however, is only a fuppofition; and your experiment

periment of reftoring electricity to a negatively electrified perfon, by extending his arm upwards into the air, with a needle between his fingers, on the point of which light may be feen in the night, is, indeed, a curious one. In this town the air is generally moifter than with us, and here I have feen Mr. *Canton* electrify the air in one room pofitively, and in another, which communicated by a door, he has electrifed the air negatively. The difference was eafily difcovered by his cork balls, as he paffed out of one room into another.——*Pere Beccaria*, too, has a pretty experiment, which fhews that air may be electrifed. Sufpending a pair of fmall light balls, by flaxen threads, to the end of his prime conductor, he turns his globe fome time, electrifing pofitively, the balls diverging and continuing feparate all the time. Then he prefents the point of a needle to his conductor, which gradually drawing off the electric fluid, the balls approach each other, and touch, before all is drawn from the conductor; opening again as more is drawn off, and feparating nearly as wide as at firft, when the conductor is reduced to the natural ftate. By this it appears, that when the balls came together, the air furrounding the balls was juft as much electrifed as the conductor at that time; and more than the conductor, when that was reduced to its natural ftate. For the balls, though in the natural ftate, will diverge, when the air that furrounds them is electrifed *plus* or *minus*, as well as when that is in its natural ftate and they are electrifed *plus* or *minus* themfelves.

I forefee

I forefee that you will apply this experiment to the fupport of your hypothefis, and I think. you may make a good
deal of it.

It was a curious enquiry of yours, Whether the electricity of the air, in clear dry weather, be of the fame denfity
at the height of two or three hundred yards, as near the
furface of the earth ; and I am glad you made the experiment.   Upon reflection, it fhould feem probable, that
whether the general ftate of the atmofphere at any time be
pofitive or negative, that part of it which is next the earth
will be nearer the natural ftate, by having given to the
earth in one cafe, or having received from it in the other.
In electrifing the air of a room, that which is neareft the
walls, or floor, is leaft altered.   There is only one fmall
ambiguity in the experiment, which may be cleared by
more trials; it arifes from the fuppofition that bodies may
be electrifed pofitively by the friction of air blowing ftrongly
on them, as it does on the kite and its ftring.   If at fome
times the electricity appears to be negative, as that friction
is the fame, the effect muft be from a negative ftate of the
upper air.

I am much pleafed with your electrical thermometer,
and the experiments you have made with it.   I formerly
fatisfied myfelf by an experiment with my phial and fy-
phon, that the elafticity of the air was not increafed by the
mere exiftence of an electric atmofphere within the phial ;
but I did not know, till you now inform me, that heat may
                                                                    be

be given to it by an electric explosion. The continuance of its rarefaction, for some time after the discharge of your glass jar and of your case of bottles, seem to make this clear. The other experiments on wet paper, wet thread, green grass, and green wood, are not so satisfactory; as possibly the reducing part of the moisture to vapour, by the electric fluid passing through it, might occasion some expansion which would be gradually reduced by the condensation of such vapour. The fine silver thread, the very small brass wire, and the strip of gilt paper, are also subject to a similar objection, as even metals, in such circumstances, are often partly reduced to smoke, particularly the gilding on paper.

But your subsequent beautiful experiment on the wire, which you made hot by the electric explosion, and in that state fired gunpowder with it, puts it out of all question, that heat is produced by our artificial electricity, and that the melting of metals in that way, is not by what I formerly called a cold fusion. A late instance here, of the melting a bell-wire, in a house struck by lightning, and parts of the wire burning holes in the floor on which they fell, has proved the same with regard to the electricity of nature. I was too easily led into that error by accounts given, even in philosophical books, and from remote ages downwards, of melting money in purses, swords in scabbards, &c. without burning the inflammable matters that were so near those melted metals. But men are, in gene-

H h h ral,

ral, such careless obfervers, that a philofopher cannot be too much on his guard in crediting their relations of things extraordinary, and fhould never build an hypothefis on any thing but clear facts and experiments, or it will be in danger of foon falling, as this does, like a houfe of cards.

How many ways there are of kindling fire, or producing heat in bodies! By the fun's rays, by collifion, by friction, by hammering, by putrefaction, by fermentation, by mixtures of fluids, by mixtures of folids with fluids, and by electricity. And yet the fire when produced, though in different bodies it may differ in circumftances, as in colour, vehemence, &c. yet in the fame bodies is generally the fame. Does not this feem to indicate that the fire existed in the body, though in a quiefcent ftate, before it was by any of thefe means excited, difengaged, and brought forth to action and to view? May it not conftitute part, and even a principal part, of the folid fubftance of bodies? If this fhould be the cafe, kindling fire in a body would be nothing more than developing this inflammable principle, and fetting it at liberty to act in feparating the parts of that body, which then exhibits the appearances of fcorching, melting, burning, &c. When a man lights an hundred candles from the flame of one, without diminifhing that flame, can it be properly faid to have *communicated* all that fire? When a fingle fpark from a flint, applied to a magazine of gunpowder, is immediately attended with this confequence, that the whole is in flame, exploding with

immenfe

immenfe violence, could all this fire exift firft in the fpark? We cannot conceive it. And thus we feem led to this fuppofition, that there is fire enough in all bodies to finge, melt, or burn them, whenever it is, by any means, fet at liberty, fo that it may exert itfelf upon them, or be difengaged from them. This liberty feems to be afforded it by the paffage of electricity through them, which we know can and does, of itfelf, feparate the parts even of water; and perhaps the immediate appearances of fire are only the effects of fuch feparations? If fo, there would be no need of fuppofing that the electric fluid *heats itfelf* by the fwiftnefs of its motion, or heats bodies by the refiftance it meets with in paffing through them. They would only be heated in proportion as fuch feparation could be more eafily made. Thus a melting heat cannot be given to a large wire in the flame of a candle, though it may to a fmall one; and this not becaufe the large wire refifts *lefs* that action of the flame which tends to feparate its parts, but becaufe it refifts it *more* than the fmaller wire; or becaufe the force being divided among more parts, acts weaker on each.

This reminds me, however, of a little experiment I have frequently made, that fhews, at one operation, the different effects of the fame quantity of electric fluid paffing through different quantities of metal. A ftrip of tinfoil, three inches long, a quarter of an inch wide at one end, and tapering all the way to a fharp point at the other, fixed between two pieces of glafs, and having the electricity of a large

glafs

glafs jar fent through it, will not be difcompofed in the broadeft part; towards the middle will appear melted in fpots; where narrower, it will be quite melted; and about half an inch of it next the point will be reduced to fmoke.

You were not miftaken in fuppofing that your account of the effect of the pointed rod, in fecuring Mr. *Weft*'s houfe from damage by a ftroke of lightning, would give me great pleafure. I thank you for it moft heartily, and for the pains you have taken in giving me fo complete a defcription of its fituation, form, and fubftance, with the draft of the melted point. There is one circumftance, *viz.* that the lightning was feen to diffufe itfelf from the foot of the rod over the wet pavement, which feems, I think, to indicate, that the earth under the pavement was very dry, and that the rod fhould have been funk deeper, till it came to earth moifter and therefore apter to receive and diffipate the electric fluid. And although, in this inftance, a conductor formed of nail rods, not much above a quarter of an inch thick, ferved well to convey the lightning, yet fome accounts I have feen from *Carolina*, give reafon to think, that larger may be fometimes neceffary, at leaft for the fecurity of the conductor itfelf, which, when too fmall, may be deftroyed in executing its office, though it does, at the fame time, preferve the houfe. Indeed, in the conftruction of an inftrument fo new, and of which we could have fo little experience, it is rather lucky that we fhould at firft be fo near the truth as we feem to be, and commit fo few errors.   There

There is another reafon for finking deeper the lower end of the rod, and alfo for turning it outwards under ground to fome diftance from the foundation; it is this, that water dripping from the eaves falls near the foundation, and fometimes foaks down there in greater quantities, fo as to come near the end of the rod though the ground about it be drier. In fuch cafe, this water may be exploded, that is, blown into vapour, whereby a force is generated that may damage the foundation. Water reduced to vapour, is faid to occupy 14,000 times its former fpace.—I have fent a charge through a fmall glafs tube, that has borne it well while empty, but when filled firft with water, was fhattered to pieces and driven all about the room:—Finding no part of the water on the table, I fufpected it to have been reduced to vapour; and was confirmed in that fufpicion afterwards, when I had filled a like piece of tube with ink, and laid it on a fheet of clean paper, whereon, after the explofion, I could find neither any moifture nor any fully from the ink. This experiment of the explofion of water, which I believe was firft made by that moft in-genious electrician father *Beccaria*, may account for what we fometimes fee in a tree ftruck by lightning, when part of it is reduced to fine fplinters like a broom; the fap veffels being fo many tubes containing a watry fluid, which when reduced to vapour, rends every tube length-ways. And perhaps it is this rarefaction of the fluids in animal bodies killed by lightning or electricity, that by

feparating

separating its fibres, renders the flesh so tender, and apt so much sooner to putrify. I think too, that much of the damage done by lightning to stone and brick walls, may sometimes be owing to the explosion of water, found, during showers, running or lodging in the joints or small cavities or cracks that happen to be in the walls.

Here are some electricians that recommend knobs instead of points on the upper end of the rods, from a supposition that the points invite the stroke. It is true that points draw electricity at greater distances in the gradual silent way; — but knobs will draw at the greatest distance a stroke. There is an experiment that will settle this. Take a crooked wire of the thickness of a quill, and of such a length as that one end of it being applied to the lower part of a charged bottle, the upper may be brought near the ball on the top of the wire that is in the bottle. Let one end of this wire be furnished with a knob, and the other be gradually tapered to a fine point. When the point is presented to discharge the bottle it must be brought much nearer before it will receive the stroke, than the knob requires to be. Points besides tend to repel the fragments of an electrised cloud, knobs draw them nearer. An experiment which I believe I have shewn you, of cotton fleece hanging from an electrised body, shows this clearly when a point or a knob is presented under it.

You seem to think highly of the importance of this discovery, as do many others on our side of the water.

Here

Here it is very little regarded; fo little, that though it is now feven or eight years fince it was made publick, I have not heard of a fingle houfe as yet attempted to be fecured by it. It is true the mifchiefs done by lightning are not fo frequent here as with us, and thofe who calculate chances may perhaps find that not one death (or the deftruction of one houfe) in a hundred thoufand happens from that caufe, and that therefore it is fcarce worth while to be at any expence to guard againft it.—But in all countries there are particular fituations of buildings more expofed than others to fuch accidents, and there are minds fo ftrongly impreffed with the apprehenfion of them, as to be very unhappy every time a little thunder is within their hearing;—it may therefore be well to render this little piece of new knowledge as general and as well underftood as poffible, fince to make us *fafe* is not all its advantage, it is fome to make us *eafy*. And as the ftroke it fecures us from might have chanced perhaps but once in our lives, while it may relieve us a hundred times from thofe painful apprehenfions, the latter may poffibly on the whole contribute more to the happinefs of mankind than the former.

Your kind wifhes and congratulations are very obliging. I return them cordially;—being with great regard and efteem,

My dear Sir,
Your affectionate friend,
and moft obedient humble fervant,
B. F.

LETTER

LETTER XXXIX.

Accounts from *Carolina (mention'd in the forego-ing Letter)* of the effects of Lightning, on two of the Rods commonly affix'd to Houses there, for securing them against Lightning.

Charles-town, Nov. 1, 1760.

" ———It is some Years since Mr. *Raven*'s Rod was struck by lightning. I hear an account of it was published at the time, but I cannot find it. According to the best information I can now get, he had fix'd to the outside of his chimney a large iron Rod, several feet in length, reach-ing above the chimney; and to the top of this rod the points were fixed. From the lower end of this rod, a small brass wire was continued down to the top of another iron rod driven into the earth. On the ground-floor in the chimney stood a gun, leaning against the back wall, nearly opposite to where the brass wire came down on the outside. The lightning fell upon the points, did no damage to the rod they were fix'd to; but the brass wire, all down till it came opposite to the top of the gun-barrel, was destroyed.*

There

A proof that it was not of sufficient substance to conduct with safety to itself (tho' with safety *so far* to the wall) so large a quantity of the electric fluid.

There the lightning made a hole through the wall or back of the chimney, to get to the gun-barrel,✝ down which it feems to have pafs'd, as, although it did not hurt the barrel, it damaged the butt of the ftock, and blew up fome bricks of the hearth. The brafs wire below the hole in the wall remain'd good.—No other damage, as I can learn, was done to the houfe.—I am told the fame houfe had formerly been ftruck by lightning, and much damaged, before thefe rods were invented."————

LETTER XL.

Mr. *William Maine's* Account of the Effects of Lightning on his Rod, dated at *Indian Land,* in *South Carolina, Aug.* 28, 1760.

———— " I had a fet of electrical points, confifting of three prongs, of large brafs wire tipt with filver, and per-fectly fharp, each about feven inches long; thefe were riveted at equal diftances into an iron nut about three quarters of an inch fquare, and opened at top equally to the diftance of fix or feven inches from point to point, in a regular triangle. This nut was fcrewed very tight on the top of an iron rod of above half an inch diameter, or the thicknefs of a common

I i i

curtain

✝ A more fubftantial conductor.

curtain rod, compofed of feveral joints, annexed by hooks turned at the ends of each joint, and the whole fixed to the chimney of my houfe by iron ftaples. The points were elevated *(a)*, fix or feven inches above the top of the chimney; and the lower joint funk three feet in the earth, in a perpendicular direction.

Thus ftood the points on Tuefday laft about five in the evening, when the lightning broke with a violent explofion on the chimney, cut the rod fquare off juft under the nutt, and I am perfuaded, melted the points, nut, and top of the rod, entirely up; as after the moft diligent fearch, nothing of either was found *(b)*, and the top of the remaining rod was cafed over with a congealed folder. The lightning ran down the rod, ftarting almoft all the ftaples *(c)*, and unhooking the joints, without affecting the rod *(d)*, except on the infide of each hook where the joints were coupled, the furface of which was melted *(e)*, and left as cafed over with folder.—No part of the chimney was damaged *(f)*, only at the foundation *(g)*, where it was fhattered almoft quite round, and feveral bricks were torn out *(b)*. Confiderable cavities were made in the earth quite round the foundation, but moft within eight or nine inches of the rod. It alfo fhattered the bottom weather-board *(i)*, at one corner of the houfe, and made a large hole in the earth by the corner poft. On the other fide of the chimney, it ploughed up feveral furrows in the earth, fome yards in length. It ran down the infide of the chimney *(k)*, carrying only foot

with

with it; and filled the whole houfe with its flafh (*l*), fmoke, and duft. It tore up the hearth in feveral places *(m)*, and broke fome pieces of china in the beaufet *(n)*. A copper tea kettle ftanding in the chimney was beat together, as if fome great weight had fallen upon it *(o)*; and three holes, each about half an inch diameter, melted through the bottom *(p)*. What feems to me moft furprifing is, that the hearth under the kettle was not hurt, yet the bottom of the kettle was drove inward, as if the lightning proceeded from under it upwards *(q)*, and the cover was thrown to the middle of the floor *(r)*. The fire dogs, an iron loggerhead, an Indian pot, an earthen cup, and a cat, were all in the chimney at the time unhurt, though great part of the hearth was torn up *(f)*. My wife's fifter, two children, and a Negro wench, were all who happened to be in the houfe at the time: The firft, and one child, fat within five feet of the chimney; and were fo ftunned, that they never faw the lightning nor heard the explofion; the wench, with the other child in her arms, fitting at a greater diftance, was fenfible of both; though every one was fo ftunn'd that they did not recover for fome time; however it pleafed God that no farther mifchief enfued. The kitchen, at 90 feet diftance, was full of Negroes, who were all fenfible of the fhock; and fome of them tell me, that they felt the rod about a minute after, when it was fo hot that they could not bear it in hand."

REMARKS.

REMARKS.

The foregoing very fenfible and diftinct account may afford a good deal of inftruction relating to the nature and effects of lightning, and to the conftruction and ufe of this inftrument for averting the mifchiefs of it.——Like other new inftruments, this appears to have been at firft in fome refpects imperfect; and we find that we are, in this as in others, to expect improvement from experience chiefly: But there feems to be nothing in the account, that fhould difcourage us in the ufe of it; fince at the fame time that its imperfections are difcovered, the means of removing them are pretty eafily to be learnt from the circumftances of the account itfelf; and its utility upon the whole is manifeft.

One intention of the pointed rod, is, to *prevent* a ftroke of lightning. *(See pages* 126, 162.) But to have a better chance of obtaining this end, the points fhould not be too near to the top of the chimney or higheft part of the building to which they are affixed, but fhould be extended five or fix feet above it; otherwife their operation in filently drawing off the fire (from fuch fragments of cloud as float in the air between the great body of cloud and the earth) will be prevented. For the experiment with the lock of cotton hanging below the electrified prime conductor, fhews, that a finger under it, being a blunt body, extends the cotton, drawing its lower part downwards; when a needle with its point prefented to the cotton, makes it fly up again to the prime conductor; and that this effect is ftrongeft, when as much

of

of the needle as poffible appears above the end of the finger; grows weaker as the needle is fhortened between the
finger and thumb; and is reduced to nothing when only a
fhort part below the point appears above the finger.   Now
it feems the points of Mr. *Maine*'s rod were elevated only
(a) fix or feven inches above the top of the chimney; which,
confidering the bulk of the chimney and the houfe, was too
fmall an elevation.   For the great body of matter near
them would hinder their being eafily brought into a negative
ftate by the repulfive power of the electrifed cloud, in
which negative ftate it is that they attract moft ftrongly
and copioufly the electric fluid from other bodies, and convey it into the earth.

(b) *Nothing of the points, &c. could be found.*   This is a
common effect. *(See page* 163.)   Where the quantity of
the electric fluid paffing is too great for the conductor thro'
which it paffes, the metal is either melted, or reduced to
fmoke and diffipated; but where the conductor is fufficiently
large, the fluid paffes in it without hurting it.   Thus thefe
three wires were deftroyed, while the rod to which they
were fixed, being of greater fubftance, remained unhurt;
its end only, to which they were joined, being a little melted, fome of the melted part of the lower ends of thofe wires
uniting with it, and appearing on it like folder.

(c) (d) (e) As the feveral parts of the rod were connected only by the ends being bent round into hooks, the
contact between hook and hook was much fmaller than
the

the rod; therefore the current through the metal being confin'd in thofe narrow paffages, melted part of the metal, as appeared on examining the infide of each hook. Where metal is melted by lightning, fome part of it is generally exploded; and thefe explofions in the joints appear to have been the caufe of unhooking them; and, by that violent action, of ftarting alfo moft of the ftaples. We learn from hence, that a rod in one continued piece is preferable to one compofed of links or parts hooked together.

(f) No part of the chimney was damaged; becaufe the lightning paffed in the rod. And this inftance agrees with others in fhewing, that the fecond and principal intention of the rods is obtainable, viz. that of *conducting* the lightning. In all the inftances yet known of the lightning's falling on any houfe guarded by rods, it has pitched down upon the point of the rod; and has not fallen upon any other part of the houfe. Had the lightning fallen on this chimney, unfurnifhed with a rod, it would probably have rent it from top to bottom, as we fee, by the effects of the lightning on the points and rod, that its quantity was very great; and we know that many chimneys have been fo demolifhed. But *no part of this was damaged, only (f) (g) (h) at the foundation, where it was fhattered and feveral bricks torn out.* Here we learn the principal defect in fixing this rod. The lower joint being funk but three feet into the earth, did not it feems go low enough to come at water, or a large body of earth fo moift as to receive readily

dily

dily from its end the quantity it conducted. The electric fluid therefore thus accumulated near the lower end of the rod, quitted it at the furface of the earth, dividing in fearch of other paffages. Part of it tore up the furface in furrows, and made holes in it: Part entered the bricks of the foundation, which being near the earth are generally moift, and, in exploding that moifture, fhattered them. *(See page* 415.) Part went through or under the foundation, and got under the hearth, blowing up great part of the bricks *(m) (J)*, and producing the other effects *(o) (p) (q) (r)*. The iron dogs, loggerhead and iron pot were not hurt, being of fufficient fubftance, and they probably protected the cat. The copper tea kettle being thin, fuffered fome damage. Perhaps, tho' found on a found part of the hearth it might at the time of the ftroke have ftood on the part blown up, which will account both for the bruifing and melting.

That *it ran down the infide of the chimney (k)* I apprehend muft be a miftake. Had it done fo, I imagine it would have brought fomething more than foot with it; it would probably have ripp'd off the pargetting, and brought down fragments of plaifter and bricks. The fhake, from the explofion on the rod, was fufficient to fhake down a good deal of loofe foot. Lightning does not ufually enter houfes by the doors, windows, or chimneys, as open paffages, in the manner that air enters them: Its nature is, to be attracted by fubftances, that are conductors of electricity; it penetrates and paffes *in* them, and, if they are not good conductors, as are nei-

ther

ther wood, brick, ftone nor plaifter, it is apt to rend them in its paffage. It would not eafily pafs thro' the air from a cloud to a building, were it not for the aid afforded it in its paffage by intervening fragments of clouds below the main body or by the falling rain.

It is faid that *the houfe was filled with its flafh (1)*. Expreffions like this are common in accounts of the effects of lightning, from which we are apt to underftand that the lightning filled the houfe. Our language indeed feems to want a word to exprefs the *light* of lightning as diftinct from the lightning itfelf. When a tree on a hill is ftruck by it, the lightning of that ftroke exifts only in a narrow vein between the cloud and tree, but its light fills a vaft fpace many miles round; and people at the greateft diftance from it are apt to fay, " the lightning came into our rooms through our windows." As it is in itfelf extreamly bright, it cannot, when fo near as to ftrike a houfe, fail illuminating highly every room in it through the windows; and this I fuppofe to have been the cafe at Mr. *Maine*'s; and that, except in and near the hearth, from the caufes abovementioned, it was not in any other part of the houfe; *the flafh* meaning no more than *the light* of the lightning.— It is for want of confidering this difference, that people fuppofe there is a kind of lightning not attended with thunder. In fact there is probably a loud explofion accompanying every flafh of lightning, and at the fame inftant;— but as found travels flower than light, we often hear the

found

found fome feconds of time after having feen the light; and as found does not travel fo far as light, we fometimes fee the light at a diftance too great to hear the found.

(*n*) The *breaking fome pieces of china in the beaufet*, may neverthelefs feem to indicate that the lightning was there: But as there is no mention of its having hurt any part of the beaufet, or of the walls of the houfe, I fhould rather afcribe that effect to the concuffion of the air, or fhake of the houfe by the explofion.

Thus, to me it appears, that the houfe and its inhabitants were faved by the rod, though the rod itfelf was unjointed by the ftroke; and that, if it had been made of one piece, and funk deeper in the earth, or had entered the earth at a greater diftance from the foundation, the mentioned fmall damages (except the melting of the points) would not have happened.

LETTER XLI.

Saturday, July 3, 1762.

TO try, at the requeft of a friend, whether amber finely powdered might be melted and run together again by means of the electric fluid, I took a piece of fmall glafs tube about $2\frac{1}{4}$ inches long, the bore about $\frac{1}{12}$ of an inch diameter, the glafs itfelf about the fame thicknefs; I introduced into this tube fome powder of amber, and

K k k

with

with two pieces of wire nearly fitting the bore, one inferted at one end, the other at the other, I rammed the
powder hard between them in the middle of the tube,
where it ftuck faft, and was in length about half an inch.
Then leaving the wires in the tube, I made them part of
the electric circuit, and difcharged through them three
rows of my cafe of bottles.   The event was, that the
glafs was broke into very fmall pieces and thofe difperfed
with violence in all directions.   As I did not expect this,
I had not, as in other experiments, laid thick paper over
the glafs to fave my eyes, fo feveral of the pieces ftruck
my face fmartly, and one of them cut my lip a little fo as
to make it bleed.   I could find no part of the amber; but
the table where the tube lay was ftained very black in
fpots, fuch as might be made by a thick fmoke forced on
it by a blaft, and the air was filled with a ftrong fmell,
fomewhat like that from burnt gunpowder.   Whence I
imagined, that the amber was burnt, and had exploded as
gunpowder would have done in the fame circumftances.

That I might better fee the effect on the amber, I made
the next experiment in a tube formed·of a card rolled up
and bound ftrongly with packthread.   Its bore was about
⅛ of an inch diameter.   I rammed powder of amber into
this as I had done in the other, and as the quantity of
amber was greater, I increafed the quantity of electric
fluid, by difcharging through it at once 5 rows of my
bottles.   On opening the tube, I found that fome of the
powder

powder had exploded, an impreffion was made on the tube though it was not burft, and moft of the powder remaining was turned black, which I fuppofe might be by the fmoke forced through it from the burnt part : Some of it was hard; but as it powdered again when preffed by the fingers, I fuppofe that hardnefs not to arife from melting any parts in it, but merely from my ramming the powder when I charged the tube.

B. F.

L E T T E R   XLII.

To the Rev. Father BECCARIA.

Rev. S I R,   *London, July* 13, 1762.

I Once promifed myfelf the pleafure of feeing you at *Turin,* but as that is not now likely to happen, being juft about returning to my native country, *America,* I fit down to take leave of you (among others of my *European* friends that I cannot fee) by writing.

I thank you for the honourable mention you have fo frequently made of me in your letters to Mr. *Collinfon* and others, for the generous defence you undertook and exe-

K k k 2   cuted

cuted with fo much fuccefs, of my electrical opinions ; and for the valuable prefent you have made me of your new work, from which I have received great information and pleafure. I wifh I could in return entertain you with any thing new of mine on that fubject ; but I have not lately purfued it. Nor do I know of any one here that is at prefent much engaged in it.

Perhaps, however, it may be agreeable to you, as you live in a mufical country, to have an account of the new inftrument lately added here to the great number that charming fcience was before poffeffed of :——As it is an inftrument that feems peculiarly adapted to *Italian* mufic, efpecially that of the foft and plaintive kind, I will endeavour to give you fuch a defcription of it, and of the manner of conftructing it, that you, or any of your friends, may be enabled to imitate it, if you incline fo to do, without being at the expence and trouble of the many experiments I have made in endeavouring to bring it to its prefent perfection.

You have doubtlefs heard the fweet tone that is drawn from a drinking glafs, by paffing a wet finger round its brim. One Mr. *Puckeridge*, a gentleman from *Ireland*, was the firft who thought of playing tunes, formed of thefe tones. He collected a number of glaffes of different fizes, fixed them near each other on a table, and tuned them by putting into them water, more or lefs, as each note required. The tones were brought out by paffing his fingers round

their

their brims.---He was unfortunately burnt here, with his inſtrument, in a fire which conſumed the houſe he lived in. Mr. *E. Delaval*, a moſt ingenious member of our Royal Society, made one in imitation of it, with a better choice and form of glaſſes, which was the firſt I ſaw or heard. Being charmed with the ſweetneſs of its tones, and the muſic he produced from it, I wiſhed only to ſee the glaſſes diſpoſed in a more convenient form, and brought together in a narrower compaſs, ſo as to admit of a greater number of tones, and all within reach of hand to a perſon ſitting before the inſtrument, which I accompliſhed, after various intermediate trials, and leſs commodious forms, both of glaſſes and conſtruction, in the following manner.

 The glaſſes are blown as near as poſ-ſible in the form of hemiſpheres, hav-ing each an open neck or ſocket in the middle. The thickneſs of the glaſs near the brim about a tenth of an inch, or hardly quite ſo much, but thicker as it comes nearer the neck, which in the largeſt glaſſes is about an inch deep, and an inch and half wide within, theſe dimenſions leſſening as the glaſſes themſelves diminiſh in ſize, except that the neck of the ſmalleſt ought not to be ſhorter than half an inch. — The largeſt glaſs is nine inches diameter, and the ſmalleſt three inches. Between theſe there are twenty-three different ſizes, differing from each other a quarter of an inch in diameter. — To make

a single instrument there should be at least six glasses blown of each size; and out of this number one may probably pick 37 glasses, (which are sufficient for 3 octaves with all the semitones) that will be each either the note one wants or a little sharper than that note, and all fitting so well into each other as to taper pretty regularly from the largest to the smallest. It is true there are not 37 sizes, but it often happens that two of the same size differ a note or half note in tone, by reason of a difference in thickness, and these may be placed one in the other without sensibly hurting the regularity of the taper form.

The glasses being chosen and every one marked with a diamond the note you intend it for, they are to be tuned by diminishing the thickness of those that are too sharp. This is done by grinding them round from the neck towards the brim, the breadth of one or two inches as may be required; often trying the glass by a well tuned harpsichord, comparing the tone drawn from the glass by your finger, with the note you want, as sounded by that string of the harpsichord. When you come near the matter, be careful to wipe the glass clean and dry before each trial, because the tone is something flatter when the glass is wet, than it will be when dry; — and grinding a very little between each trial, you will thereby tune to great exactness. The more care is necessary in this, because if you go below your required tone, there is no sharpening it again

but

but by grinding fomewhat off the brim, which will afterwards require polifhing, and thus encreafe the trouble.

The glaffes being thus tuned, you are to be provided with a cafe for them, and a fpindle on which they are to be fixed. My cafe is about three feet long, eleven inches every way wide within at the biggeft end, and five inches at the fmalleft end; for it tapers all the way, to adapt it better to the conical figure of the fet of glaffes. This cafe opens in the middle of its height, and the upper part turns up by hinges fixed behind. The fpindle which is of hard iron, lies horizontally from end to end of the box within, exactly in the middle, and is made to turn on brafs gudgeons at each end. It is round, an inch diameter at the thickeft end, and tapering to a quarter of an inch at the fmalleft. — A fquare fhank comes from its thickeft end through the box, on which fhank a wheel is fixed by a fcrew. This wheel ferves as a fly to make the motion equable, when the fpindle, with the glaffes, is turned by the foot like a fpinning wheel. My wheel is of mahogany, 18 inches diameter, and pretty thick, fo as to conceal near its circumference about 25 lb of lead. — An ivory pin is fixed in the face of this wheel and about 4 inches from the axis. Over the neck of this pin is put the loop of the ftring that comes up from the moveable ftep to give it motion. The cafe ftands on a neat frame with four legs.

To fix the glaffes on the fpindle, a cork is firft to be fitted in each neck pretty tight, and projecting a little

without

without the neck, that the neck of one may not touch the
infide of another when put together, for that would make
a jarring. — Thefe corks are to be perforated with holes
of different diameters, fo as to fuit that part of the fpindle
on which they are to be fixed.   When a glafs is put on,
by holding it ftiffly between both hands, while another
turns the fpindle, it may be gradually brought to its place.
But care muft be taken that the hole be not too fmall,
left in forcing it up the neck fhould fplit; nor too large,
left the glafs not being firmly fixed, fhould turn or move
on the fpindle, fo as to touch and jar againft its neighbouring glafs.   The glaffes thus are placed one in another, the
largeft on the biggeft end of the fpindle which is to the
left hand; the neck of this glafs is towards the wheel, and
the next goes into it in the fame pofition, only about an
inch of its brim appearing beyond the brim of the firft;
thus proceeding, every glafs when fixed fhows about an
inch of its brim, (or three quarters of an inch, or half an
inch, as they grow fmaller) beyond the brim of the glafs
that contains it; and it is from thefe expofed parts of each
glafs that the tone is drawn, by laying a finger upon one of
them as the fpindle and glaffes turn round.

My largeft glafs is G a little below the reach of a common voice, and my higheft G, including three compleat
octaves. —— To diftinguifh the glaffes the more readily to
the eye, I have painted the apparent parts of the glaffes
within fide, every femitone white, and the other notes of
the

the octave with the feven prifmatic colours, *viz.* C, red; D, orange; E, yellow; F, green; G, blue; A, Indigo; B, purple; and C, red again; — fo that glaffes of the fame colour (the white excepted) are always octaves to each other.

This inftrument is played upon, by fitting before the middle of the fet of glaffes as before the keys of a harpfichord, turning them with the foot, and wetting them now and then with a fpunge and clean water. The fingers fhould be firft a little foaked in water and quite free from all greafinefs; a little fine chalk upon them is fometimes ufeful, to make them catch the glafs and bring out the tone more readily. Both hands are ufed, by which means different parts are played together.---Obferve, that the tones are beft drawn out when the glaffes turn *from* the ends of the fingers, not when they turn *to* them.

The advantages of this inftrument are, that its tones are incomparably fweet beyond thofe of any other; that they may be fwelled and foftened at pleafure by ftronger or weaker preffures of the finger, and continued to any length; and that the inftrument, being once well tuned, never again wants tuning.

In honour of your mufical language, I have borrowed from it the name of this inftrument, calling it the *Armonica.*

With great efteem and refpect, I am, &c.

L l l   L E T-

LETTER XLIII.

From Profeſſor WINTHROP, to B. F.

S I R,   *Cambridge, N. E. Sept.* 29, 1762.

THERE is an obſervation relating to electricity in the atmoſphere, which ſeemed new to me, though perhaps it will not to you : However, I will venture to mention it. I have ſome points on the top of my houſe, and the wire where it paſſes within-ſide the houſe is furniſhed with bells, according to your method, to give notice of the paſſage of the electric fluid. In ſummer, theſe bells generally ring at the approach of a thunder cloud ; but ceaſe ſoon after it begins to rain. In winter, they ſometimes, though not very often, ring while it is ſnowing ; but. never, that I remember, when it rains. But what was unexpected to me was, that, though the bells had not rung while it was ſnowing, yet the next day, after it had done ſnowing, and the weather was cleared up ; while the ſnow was driven about by a high wind at W. or N. W. the bells rung for ſeveral hours (though with little intermiſſions) as briſkly as ever I knew them, and I drew conſiderable ſparks from the wire. This phænomenon I never obſerved but twice ; *viz.* on the 31ſt of *January*, 1760, and the 3d of *March*, 1762.

I am, Sir, &c.

LE T-

LETTER XLIV.

To a Friend.

Dear SIR,   *July* 20, 176ẓ.

I Have perufed your paper on found, and would freely mention to you, as you defire it, every thing that appeared to me to need correction :— But nothing of that kind occurs to me, unlefs it be, where you fpeak of the air as " the *beft* medium for conveying found." Perhaps this is fpeaking rather too pofitively, if there be, as I think there are, fome other mediums that will convey it farther and more readily. — It is a well-known experiment, that the fcratching of a pin at one end of a long piece of timber, may be heard by an ear applied near the other end, though it could not be heard at the fame diftance through the air. —— And two ftones being ftruck fmartly together under water, the ftroke may be heard at a greater diftance by an ear alfo placed under water in the fame river, than it can be heard through the air. I think I have heard it near a mile; how much farther it may be heard, I know not; but fuppofe a great deal farther, becaufe the found did not feem faint, as if at a diftance, like diftant founds through air, but fmart and ftrong, and as if prefent juft at the ear.——I wifh you would repeat thefe

L l l 2   expe-

experiments now you are upon the fubject, and add your own obfervations. —— And if you were to repeat, with your naturally exact attention and obfervation, the common experiment of the bell in the exhaufted receiver, poffibly fomething new may occur to you, in confidering,

1. Whether the experiment is not ambiguous; *i. e.* whether the gradual exhaufting of the air, as it creates an increafing difference of preffure on the outfide, may not occafion in the glafs a difficulty of vibrating, that renders it lefs fit to communicate to the air without, the vibrations that ftrike it from within; and the diminution of the found arife from this caufe, rather than from the diminution of the air?

2. Whether as the particles of air themfelves are at a diftance from each other, there muft not be fome medium between them, proper for conveying found, fince otherwife it would ftop at the firft particle?

3. Whether the great difference we experience in hearing founds at a diftance, when the wind blows towards us from the fonorous body, or towards that from us, can be well accounted for by adding to or fubftracting from the fwiftnefs of found, the degree of fwiftnefs that is in the wind at the time? The latter is fo fmall in proportion, that it feems as if it could fcarce produce any fenfible effect, and yet the difference is very great. Does not this give fome hint, as if there might be a fubtile fluid, the conductor of found, which moves at different times in different directions over the furface of the earth, and whofe

whofe motion may perhaps be much fwifter than that of the air in our ftrongeft winds; and that in paffing through air, it may communicate that motion to the air which we call wind, though a motion in no degree fo fwift as its own?

4. It is fomewhere related, that a piftol fired on the top of an exceeding high mountain, made a noife like thunder in the valleys below. Perhaps this fact is not exactly related: but if it is, would not one imagine from it, that the rarer the air, the greater found might be produced in it from the fame caufe?

5. Thofe balls of fire which are fometimes feen paffing over a country, computed by philofophers to be often 30 miles high at leaft, fometimes burft at that height; the air muft be exceeding rare there, and yet the explofion produces a found that is heard at that diftance, and for 70 miles round on the furface of the earth, fo violent too as to fhake buildings, and give an apprehenfion of an earthquake. Does not this look as if a rare atmofphere, almoft a vacuum, was no bad conductor of found?

I have not made up my own mind on thefe points, and only mention them for your confideration, knowing that every fubject is the better for your handling it.

With the greateft efteem, I am, &c.   B. F.

L E T T E R XLV.

To Dr. P. in London.

S I R,   *Philadelphia, Dec.* 1, 1762.

DURING our paſſage to Madeira, the weather being warm, and the cabbin windows conſtantly open
for the benefit of the air, the candles at night flared and
run very much, which was an inconvenience.   At Madeira
we got oil to burn, and with a common glaſs tumbler or
beaker, ſlung in wire, and ſuſpended to the cieling of the
cabbin, and a little wire hoop for the wick, furniſh'd with
corks to float on the oil, I made an Italian lamp, that gave
us very good light all over the table.—The glaſs at bottom
contained water to about one third of its height; another
third was taken up with oil; the reſt was left empty that
the ſides of the glaſs might protect the flame from the
wind.   There is nothing remarkable in all this; but what
follows is particular.   At ſupper, looking on the lamp, I remarked that tho' the ſurface of the oil was perfectly tranquil,
and duly preſerved its poſition and diſtance with regard to
the brim of the glaſs, the water under the oil was in great
commotion, riſing and falling in irregular waves, which
                                                    continued

continued during the whole evening. The lamp was kept burning as a watch light all night, till the oil was fpent, and the water only remain'd. In the morning I obferved, that though the motion of the fhip continued the fame, the water was now quiet, and its furface as tranquil as that of the oil had been the evening before. At night again, when oil was put upon it, the water refum'd its irregular motions, rifing in high waves almoft to the furface of the oil, but without difturbing the fmooth level of that furface. And this was repeated every day during the voyage.

Since my arrival in America, I have repeated the experiment frequently thus. I have put a pack-thread round a tumbler, with ftrings of the fame, from each fide, meeting above it in a knot at about a foot diftance from the top of the tumbler. Then putting in as much water as would fill about one third part of the tumbler, I lifted it up by the knot, and fwung it to and fro in the air; when the the water appeared to keep its place in the tumbler as fteadily as if it had been ice.—But pouring gently in upon the water about as much oil, and then again fwinging it in the air as before, the tranquility before poffeffed by the water, was transferred to the furface of the oil, and the water under it was agitated with the fame commotions as at fea.

I have fhewn this experiment to a number of ingenious perfons. Thofe who are but flightly acquainted with the principles of hydroftatics, &c. are apt to fancy immediately that they underftand it, and readily attempt to explain

it;

it; but their explanations have been different, and to me not very intelligible.—Others more deeply ſkill'd in thoſe principles, ſeem to wonder at it, and promiſe to conſider it. And I think it is worth conſidering: For a new appearance, if it cannot be explain'd by our old principles, may afford us new ones, of uſe perhaps in explaining ſome other obſcure parts of natural knowledge.

I am, &c.   B. F.

L E T T E R XLVI.

From Mr. A. S. to B. F.

I Have juſt recollected that in one of our great ſtorms of lightning, I ſaw an appearance, which I never obſerved before, nor ever heard deſcribed. I am perſuaded that I ſaw *the* flaſh which ſtruck St. Bride's ſteeple. Sitting at my window, and looking to the north, I ſaw what appeared to me a ſolid ſtreight rod of fire, moving at a very ſharp angle with the horizon. It appeared to my eye as about two inches diameter, and had nothing of the zig-zag lightning motion. I inſtantly told a perſon ſitting with me, that

ſome

ſome place muſt be ſtruck at that inſtant. I was ſo much ſurprized at the vivid diſtinct appearance of the fire, that I did not hear the clap of thunder, which ſtunned every one befides. Confidering how low it moved, I could not have thought it had gone ſo far, having St. Martin's, the New Church, and St. Clement's ſteeples in its way. It ſtruck the ſteeple a good way from the top, and the firſt impreſſion it made in the ſide is in the ſame direction I ſaw it move in. It was ſucceeded by two flaſhes, almoſt united, moving in a pointed direction. There were two diſtinct houſes ſtruck in Eſſex ſtreet. I ſhould have thought the rod would have fallen in Covent Garden, it was ſo low. Perhaps the appearance is frequent, though never before ſeen by

Yours,   A. S.

L E T T E R   XLVII.

To Mr. P. F. Newport.

———— You may acquaint the gentleman that deſired you to enquire my opinion of the beſt method of ſecuring a powder magazine from lightning, that I think they can-

M m m

not

not do better than to erect a maft not far from it, which m^y reach 15 or 20 feet above the top of it, with a thick iron rod in one piece faftened to it, pointed at the higheft end, and reaching down through the earth till it comes to water. Iron is a cheap metal; but if it were dearer, as this is a publick thing, the expence is infignificant; therefore I would have the rod at leaft an inch thick, to allow for its gradually wafting by ruft; it will laft as long as the maft, and may be renewed with it. The fharp point for five or fix inches fhould be gilt.

But there is another circumftance of importance to the ftrength, goodnefs and ufefulnefs of the powder, which does not feem to have been enough attended to: I mean the keeping it perfectly dry. For want of a method of doing this, much is fpoilt in damp magazines, and much fo damaged as to become of little value.—If inftead of barrels it were kept in cafes of bottles well cork'd; or in large tin canifters, with fmall covers fhutting clofe by means of oil'd paper between, or covering the joining on the canifter; or if in barrels, then the barrels lined with thin fheet lead; no moifture in either of thefe methods could poffibly enter the powder, fince glafs and metals are both impervious to water.

By the latter of thefe means you fee tea is brought dry and crifp from China to Europe, and thence to America, tho' it comes all the way by fea in the damp hold of a fhip.

And

And by this method, grain, meal, &c. if well dry'd before 'tis put up may be kept for ages found and good.

There is another thing very proper to line fmall barrels with; it is what they call tin-foil, or leaf-tin, being tin mill'd between rollers till it becomes as thin as paper, and more pliant, at the fame time that its texture is extreamly clofe. It may be apply'd to the wood with common pafte, made with boiling water thicken'd with flour; and, fo laid on, will lie very clofe and ftick well: But I fhould prefer a hard fticky varnifh for that purpofe, made of linfeed oil much boil'd. The heads might be lined feparately, the tin wrapping a little round their edges. The barrel while the lining is laid on, fhould have the end hoops flack, fo that the ftaves ftanding at a little diftance from each other, may admit the head into its groove. The tin-foil fhould be plyed into the groove. Then one head being put in, and that end hoop'd tight, the barrel would be fit to receive the powder, and when the other head is put in and the hoops drove up, the powder would be fafe from moifture even if the barrel were kept under water. This tin-foil but about 18 pence fterling a pound, and is fo extreamly thin, that I imagine a pound of it would line three or four powder barrels.

I am &c.   B. F.

LETTER XLVIII.

To Mifs S———n, at *Wanſtead*.

Craven-ſtreet, May 17, 1760.

I Send my dear good girl the books I mentioned to her laſt night. I beg her to accept them as a ſmall mark of my eſteem and friendſhip. They are written in the familiar eaſy manner for which the French are ſo remarkable, and afford a good deal of philoſophic and practical knowledge, unembarras'd with the dry mathematics uſed by more exact reaſoners, but which is apt to diſcourage young beginners.—I would adviſe you to read with a pen in your hand, and enter in a little book ſhort hints of what you find that is curious, or that may be uſeful; for this will be the beſt method of imprinting ſuch particulars in your memory, where they will be ready, either for practice on ſome future occaſion, if they are matters of utility; or at leaſt to adorn and improve your converſarion, if they are rather points of curioſity.—And, as many of the terms of ſcience are ſuch as you cannot have met with in your common reading, and may therefore be unacquainted with, I think it would be well for you to have a good dictionary at hand, to conſult immediately when you meet with a word

you

you do not comprehend the precife meaning of. This may at firft feem troublefome and interrupting; but 'tis a trouble that will daily diminifh, as you will daily find lefs and lefs occafion for your Dictionary as you become more acquainted with the terms; and in the mean time you will read with more fatisfaction becaufe with more underftanding.—When any point occurs in which you would be glad to have farther information than your book affords you, I beg you would not in the leaft apprehend that I fhould think it a trouble to receive and anfwer your queftions. It will be a pleafure, and no trouble. For though I may not be able, out of my own little ftock of knowledge to afford you what you require, I can eafily direct you to the books where it may moft readily be found. Adieu, and believe me ever, my dear friend,

Yours affectionately,
B. FRANKLIN.

L E T T E R   XLIX.

To the fame.

Craven-ftreet, June 11, 1760.

T I S a very fenfible queftion you afk, how the air can affect the barometer, when its opening appears covered with wood?—If indeed it was fo clofely covered

as

as to admit of no communication of the outward air to the surface of the mercury, the change of weight in the air could not poſſibly affect it. But the leaſt crevice is ſuffici-ent for the purpoſe ; a pinhole will do the buſineſs. And if you could look behind the frame to which your baro-meter is fixed, you would certainly find ſome ſmall opening.

There are indeed ſome barometers in which the body of mercury at the lower end is contained in a cloſe leather bag, and ſo the air cannot come into immediate contact with the mercury ; yet the ſame effect is produced. For the leather being flexible, when the bag is preſſed by any additional weight of air, it contracts, and the mercury is forced up into the tube ;—when the air becomes lighter, and its preſſure leſs, the weight of the mercury prevails, and it deſcends again into the bag.

Your obſervation on what you have lately read concern-ing inſects, is very juſt and ſolid. Superficial minds are apt to deſpiſe thoſe who make that part of the creation their ſtudy, as mere triflers ; but certainly the world has been much obliged to them. Under the care and manage-ment of man, the labours of the little Silkworm afford employment and ſubſiſtence to thouſands of families, and become an immenſe article of commerce. The Bee, too, yields us its delicious honey, and its wax uſeful to a multi-tude of purpoſes. Another inſect, it is ſaid, produces the Cochineal, from whence we have our rich ſcarlet dye.

The

The ufefulnefs of the Cantharides or Spanifh flies, in medicine, is known to all, and thoufands owe their lives to that knowledge. By human induftry and obfervation, other properties of other infects may poffibly be hereafter difcovered, and of equal utility. A thorough acquaintance with the nature of thefe little creatures, may alfo enable mankind to prevent the increafe of fuch as are noxious or fecure us againft the mifchiefs they occafion. Thefe things doubtlefs your books make mention of: I can only add a particular late inftance which I had from a Swedifh gentleman of good credit.—In the green timber intended for fhip-building at the king's yards in that country, a kind of worms were found, which every year became more numerous and more pernicious, fo that the fhips were greatly damaged before they came into ufe. The king fent Linnæus, the great naturalift, from Stockholm, to enquire into the affair, and fee if the mifchief was capable of any remedy. He found, on examination, that the worm was produced from a fmall egg, depofited in the little roughneffes on the furface of the wood, by a particular kind of fly or beetle; from whence the worm, as foon as it was hatch'd, began to eat into the fubftance of the wood, and after fome time came out again a fly of the parent kind, and fo the fpecies increas'd. The feafon in which the fly laid its eggs, Linnæus knew to be about a fortnight (I think) in the month of May, and at no other time in the year. He therefore advis'd, that fome days before that feafon, all the green timber fhould be thrown into the water, and kept under water till the feafon was over. Which being done by the king's order, the flies miffing their ufual nefts, could not increafe; and the fpecies was either deftroyed or went elfewhere ; and the wood was effectually preferved, for after the firft year, it became too dry and hard for their purpofe.

There is however, a prudent moderation to he ufed in ftudies of this kind. The knowledge of nature may be ornamental, and it may be ufeful; but if to attain an eminence in that, we neglect the knowledge and practice of effential duties, we deferve reprehenfion. For there is no rank in natural knowledge of equal dignity and importance with that of being a good parent, a good child, a good hufband, or wife, a good neighbour or friend, a good fubject or citizen, that is, in fhort, a good chriftian. Nicholas Gimcrack, therefore, who neglected the care of his family, to purfue butterflies, was a juft object of ridicule, and we muft give him up as fair game to the fatyrift.

Adieu, my dear friend, and believe me ever

Yours affectionately,

B. FRANKLIN.

L E T-

LETTER L.

To the fame.

My dear Friend,   *London Sept.* 13, 1760.

I HAVE your agreeable letter from *Briftol,* which I take this firft leifure hour to anfwer, having for fome time been much engaged in bufinefs.

Your firft queftion, *What is the reafon the water at this place, though cold at the fpring, becomes warm by pumping?* It will be moft prudent in me to forbear attempting to anfwer, till, by a more circumftantial account, you affure me of the fact. I own I fhould expect that operation to warm, not fo much the water pumped, as the perfon pumping.—The rubbing of dry folids together, has been long obferved to produce heat; but the like effect has never yet, that I have heard, been produced by the mere agitation of fluids, or friction of fluids with folids. Water in a bottle fhook for hours by a mill hopper, it is faid, difcovered no fenfible addition of heat. The production of animal heat by exercife, is therefore to be accounted for in another manner, which I may hereafter endeavour to make you acquainted with.

This prudence of not attempting to give reasons before one is sure of facts, I learnt from one of your sex, who, as *Selden* tells us, being in company with some gentlemen that were viewing, and considering something which they called a Chinese shoe, and disputing earnestly about the manner of wearing it, and how it could possibly be put on; put in her word, and said modestly, *Gentlemen, are you sure it is a shoe?—Should not that be settled first?*

But I shall now endeavour to explain what I said to you about the tide in rivers, and to that end shall make a figure, which though not very like a river, may serve to convey my meaning.—Suppose a canal 140 miles long, communi- cating at one end with the sea, and filled therefore with sea water. I chuse a canal at first, rather than a river, to throw out of consideration the effects produced by the streams of fresh water from the land, the inequality in breadth, and the crookedness of courses.

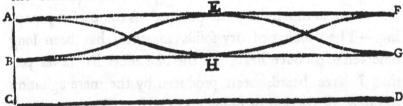

Let A, C, be the head of the canal; C, D, the bottom of it; D, F, the open mouth of it next the sea. Let the strait prick'd line, B, G, represent low water mark the whole length of the canal, A, F, high water mark:—Now if a person standing at E, and observing at the time of high water there, that the canal is quite full at that place up to the

the line E, fhould conclude that the canal is equally full to the fame height from end to end, and therefore there was as much more water come into the canal fince it was down at low water mark, as would be included in the oblong fpace A, B, G, F, he would be greatly miftaken. For the tide is *a Wave*, and the top of the wave, which makes high water, as well as every other lower part, is progreffive; and it is high water fucceffively, but not at the fame time, in all the feveral points between G, F, and A, B.—And in fuch a length as I have mentioned it is low water at F, G, and alfo at A, B, at or near the fame time with its being high water at E; fo that the furface of the water in the canal, during that fituation, is properly reprefented by the curve pricked line B, E, G. And on the other hand, when it is low water at E, H, it is high water both at F, G, and at A, B, at or near the fame time; and the furface would then be defcribed by the inverted curve line, A, H, F.

In this view of the cafe, you will eafily fee, that there muft be very little more water in the canal at what we call high water, than there is at low water, thofe terms not re-lating to the whole canal at the fame time, but fucceffively to its parts. And if you fuppofe the canal fix times as long, the cafe would not vary as to the quantity of water at different times of the tide; there would only be fix waves in the canal at the fame time, inftead of one, and the hol-lows in the water would be equal to the hills.

That

That this is not mere theory, but conformable to fact, we know by our long rivers in *America*. The *Delaware*, on which *Philadelphia* stands, is in this particular similar to the canal I have supposed of one wave: For when it is high water at the *Capes* or mouth of the river, it is also high water at *Philadelphia*, which stands about 140 miles from the sea; and there is at the same time a low water in the middle between the two high waters; where, when it comes to be high water, it is at the same time low water at the *Capes* and at *Philadelphia*. And the longer rivers have, some a wave and half, some two, three, or four waves, according to their length.—In the shorter rivers of this island, one may see the same thing in part: for instance, it is high water at *Gravesend* an hour before it is high water at *London Bridge*; and 20 miles below *Gravesend* an hour before it is high water at *Gravesend*. Therefore at the time of high water at *Gravesend* the top of the wave is there, and the water is then not so high by some feet where the top of the wave was an hour before, or where it will be an hour after, as it is just then at *Gravesend*.

Now we are not to suppose, that because the swell or top of the wave runs at the rate of 20 miles an hour, that therefore the current or water itself of which the wave is composed, runs at that rate. Far from it. To conceive this motion of a wave, make a small experiment or two. Fasten one end of a cord in a window near the top of a house, and let the other end come down to the ground; take this end in

your

your hand, and you may, by a fudden motion occafion a wave in the cord that will run quite up to the window; but though the wave is progreffive from your hand to the window, the parts of the rope do not proceed with the wave, but remain where they were, except only that kind of motion that produces the wave.—So if you throw a ftone into a pond of water when the furface is ftill and fmooth, you will fee a circular wave proceed from the ftone as its center, quite to the fides of the pond; but the water does not proceed with the wave, it only rifes and falls to form it in the different parts of its courfe; and the waves that follow the firft, all make ufe of the fame water with their predeceffors.

But a wave in water is not indeed in all circumftances exactly like that in a cord; for water being a fluid, and gravitating to the earth, it naturally runs from a higher place to a lower; therefore the parts of the wave in water do actually run a little both ways from its top towards its lower fides, which the parts of the wave in the cord cannot do. Thus when it is high and ftanding water at *Gravefend*, the water 20 miles below has been running ebb, or towards the fea for an hour, or ever fince it was high water there; but the water at *London Bridge* will run flood, or from the fea yet another hour, till it is high water or the top of the wave arrives at that bridge, and then it will have run ebb an hour at *Gravefend*, &c. &c. Now this motion of the water, occafioned only by its gravity, or tendency to run from a higher

place

place to a lower, is by no means so swift as the motion of the wave. It scarce exceeds perhaps two miles in an hour. If it went as the wave does twenty miles an hour, no ships could ride at anchor in such a stream, nor boats row against it.

In common speech, indeed, this current of the water both ways from the top of the wave is called *the tide*; thus we say, *the tide runs strong, the tide runs at the rate of one, two, or three miles an hour,* &c. and when we are at a part of the river behind the top of the wave, and find the water lower than high-water mark, and running towards the sea we say, *the tide runs ebb*; and when we are before the top of the wave, and find the water higher than low-water mark, and running from the sea, we say, the *tide runs flood*; but these expressions are only locally proper; for a tide strictly speaking, is *one whole wave*, including all its parts higher and lower, and these waves succeed one another about twice in 24 hours.

This motion of the water occasioned by its gravity, will explain to you why the water near the mouths of rivers may be salter at high water than at low. Some of the salt water, as the tide wave enters the river, runs from its top and fore side, and mixes with the fresh, and also pushes it back up the river.

Supposing that the water commonly runs during the flood at the rate of two miles in an hour, and that the flood runs five hours, you see that it can bring at most into our canal

only

only a quantity of water equal to the space included in the breadth of the canal, ten miles of its length, and the depth between low and high-water mark; which is but a fourteenth part of what would be necessary to fill all the space between low and high-water mark, for 140 miles, the whole length of the canal.

And indeed such a quantity of water as would fill that whole space, to run in and out every tide, must create so outrageous a current as would do infinite damage to the shores, shipping, &c. and make the navigation of a river almost impracticable.

I have made this letter longer than I intended, and therefore reserve for another what I have further to say on the subject of tides and rivers. I shall now only add, that I have not been exact in the numbers, because I would avoid perplexing you with minute calculations, my design at present being chiefly to give you distinct and clear ideas of the first principles.

After writing six folio pages of philosophy to a young girl, is it necessary to finish such a letter with a compliment?—Is not such a letter of itself a compliment?— Does it not say, she has a mind thirsty after knowledge, and capable of receiving it; and that the most agreeable things one can write to her are those that tend to the improvement of her understanding?—It does indeed say all this, but then it is still no compliment; it is no more than plain honest truth, which is not the character of a compliment. So if I

would

would finifh my letter in the mode, I fhould yet add fome
thing that means nothing, and is *merely* civil and polite.—
But being naturally aukward at every circumftance of cere-
mony, I fhall not attempt it. I had rather conclude abruptly with what pleafes me more than any compliment
can pleafe you, that I am allowed to fubfcribe myfelf

Your affectionate friend,

B. FRANKLIN.

LETTER LI.

To the fame.

My dear Friend,   *Craven-ftr. Monday March* 30, 1761.

SUPPOSING the fact, that the water of the well at
Briftol is warmer after fometime pumping, I think
your manner of accounting for that increafed warmth very
ingenious and probable. It did not occur to me, and therefore I doubted of the fact.

You are, I think, quite right in your opinion, that the
rifing of the tides in rivers is not owing to the immediate
influence

influence of the moon on the rivers. It is rather a fubfequent effect of the influence of the moon on the fea, and
does not make its appearance in fome rivers till the moon
has long pafs'd by. I have not exprefs'd myfelf clearly if
you have underftood me to mean otherwife. You know I
have mentioned it as a fact, that there are in fome rivers
feveral tides all exifting at the fame time; that is, two,
three, or more, high-waters, and as many low-waters, in
different parts of the fame river, which cannot poffibly be
all effects of the moon's immediate action on that river;
but they may be fubfequent effects of her action on the
fea.

In the enclofed paper you will find my fentiments on
feveral points relating to the air, and the evaporation of
water. It is Mr. Collinfon's copy, who took it from one I
fent thro' his hands to a correfpondent in *France* fome years
fince; I have, as he defired me, corrected the miftakes he
made in tranfcribing, and muft return it to him; but if you
think it worth while, you may take a copy of it: I
would have faved you any trouble of that kind, but had
not time.

Some day in the next or the following week, I purpofe
to have the pleafure of feeing you at *Wanftead*; I fhall
accompany your good mama thither, and ftay till the next
morning, if it may be done without incommoding your
family too much.—We may then difcourfe any points in
that paper that do not feem clear to you; and taking a walk

to

to lord *Tilney's* ponds, make a few experiments there to explain the nature of the tides more fully. In the mean time, believe me to be, with the higheſt eſteem and regard, your ſincerely affectionate friend,

B. FRANKLIN.

LETTER LII.

To the ſame.

Cravenſtr. Aug. 10, 1761.

WE are to ſet out this week for *Holland,* where we may poſſibly ſpend a month, but purpoſe to be at home again before the coronation. I could not go without taking leave of you by a line at leaſt, when I am ſo many letters in your debt.

In yours of *May* 19, which I have before me, you ſpeak of the eaſe with which ſalt water may be made freſh by diſtillation, ſuppoſing it to be, as I had ſaid, that in evaporation the air would take up water but not the ſalt that was

mixed

mixed with it. It is true that diftilled fea water will not be falt, but there are other difagreeable qualities that rife with the water in diftillation; which indeed feveral befides Dr. *Hales* have endeavoured by fome means to prevent; but as yet their methods have not been brought much into ufe.

I have a fingular opinion on this fubject, which I will venture to communicate to you, though I doubt you will rank it among my whims.——It is certain that the fkin has *imbibing* as well as *difcharging* pores; witnefs the effects of a bliftering plaifter, &c. I have read that a man hired by a phyfician to ftand by way of experiment in the open air naked during a moift night, weighed near three pounds heavier in the morning. I have often obferved myfelf, that however thirfty I may have been before going into the water to fwim, I am never long fo in the water. Thefe imbibing pores, however, are very fine, perhaps fine enough in filtring to feparate falt from water; for though I have foaked (by fwimming, when a boy) feveral hours in the day for feveral days fucceffively in falt-water, I never found my blood and juices falted by that means, fo as to make me thirfty or feel a falt tafte in my mouth: And it is remarkable that the flefh of fea fifh, though bred in falt water is not falt.——Hence I imagine, that if people at fea, diftreffed by thirft when their frefh water is unfortunately fpent, would make bathing-tubs of their empty water cafks, and filling them with fea water, fit in them an hour or two

each

each day, they might be greatly relieved. Perhaps keeping their cloaths conftantly wet might have an almoft e-
qual effect; and this without danger of catching cold.
Men do not catch cold by wet cloaths at fea. Damp but
not wet linen may poffibly give colds; but no one catches
cold by bathing, and no cloaths can be wetter than water
itfelf. Why damp cloaths fhould then occafion colds, is
a curious queftion, the difcuffion of which I referve for a
future letter, or fome future converfation.

Adieu, my little philofopher. Prefent my refpect-
ful compliments to the good ladies your aunts, and to mifs
Pitt; and believe me ever

Your affectionate friend,
and humble Servant,
B. FRANKLIN.

L E T T E R LIII.
To the fame.

London, March 22, 1762.

I MUST retract the charge of idlenefs in your ftudies,
when I find you have gone thro' the doubly difficult
tafk of reading fo big a book, on an abftrufe fubject and in
a foreign language.

In

In anfwer to your queftion concerning the *Leiden* phial; —The hand that holds the bottle receives and conducts away the electric fluid that is driven out of the outfide by the repulfive power of that which is forced into the infide of the bottle. As long as that power remains in the fame fituation, it muft prevent the return of what it had expelled; though the hand would readily fupply the quantity if it could be received.

Your affectionate Friend,

B. FRANKLIN.

L E T T E R LIV.

To the fame.

Craven-ftreet, Saturday Evening, paft 10.

THE queftion you afk me is a very fenfible one, and I fhall be glad if I can give you a fatisfactory anfwer. There are two ways of contracting a chimney; one, by contracting the opening *before* the fire; the other, by contracting the funnel *above* the fire. If the funnel above the fire is left open in its full dimenfions, and the opening before

tore

fore the fire is contracted; then the coals, I imagine, will burn faster, because more air is directed through the fire, and in a stronger stream; that air which before passed over it, and on each side of it, now passing *thro'* it. This is seen in narrow stove chimneys, when a sacheverell or blower is used, which still more contracts the narrow opening.—But if the funnel only *above* the fire is contracted, then, as a less stream of air is passing up the chimney, less must pass through the fire, and consequently it should seem that the consuming of the coals would rather be checked than augmented by such contraction. And this will also be the case, when both the opening *before* the fire, and the funnel *above* the fire are contracted, provided the funnel above the fire is more contracted in proportion than the opening before the fire.——So you see I think you had the best of the argument; and as you notwithstanding gave it up in complaisance to the company, I think you had also the best of the dispute. There are few, though convinced, that know how to give up, even an error, they have been once engaged in maintaining; there is therefore the more merit in dropping a contest where one thinks one's self right; 'tis at least respectful to those we converse with. And indeed all our knowledge is so imperfect, and we are from a thousand causes so perpetually subject to mistake and error, that positiveness can scarce ever become even the most knowing; and modesty in advancing any opinion, however plain and true we may suppose it, is

always

always decent, and generally more likely to procure affent.   *Pope*'s Rule

To fpeak, though fure, with feeming diffidence,

is therefore a good one; and if I had ever feen in your converfation the leaft deviation from it, I fhould earneftly recommend it to your obfervation.

I am, &c.

B. FRANKLIN.

L E T T E R LV.

To Mr. O. N.

Dear S I R,

I Cannot be of opinion with you that 'tis too late in life for you to learn to fwim.   The river near the bottom of your garden affords you a moft convenient place for the purpofe.   And as your new employment requires your being often on the water, of which you have fuch a dread, I think you would do well to make the trial; nothing being fo likely to remove thofe apprehenfions as the confcioufnefs of an ability to fwim to the fhore, in cafe of an

accident,

accident, or of supporting yourself in the water till a boat could come to take you up.

I do not know how far corks or bladders may be useful in learning to swim, having never seen much trial of them. Possibly they may be of service in supporting the body while you are learning what is called the stroke, or that manner of drawing in and striking out the hands and feet that is necessary to produce progressive motion. But you will be no swimmer till you can place some confidence in the power of the water to support you; I would therefore advise the acquiring that confidence in the first place; especially as I have known several who by a little of the practice necessary for that purpose, have insensibly acquired the stroke, taught as it were by nature.

The practice I mean is this. Chusing a place where the water deepens gradually, walk coolly into it till it is up to your breast, then turn round, your face to the shore, and throw an egg into the water between you and the shore. It will sink to the bottom, and be easily seen there, as your water is clear. It must lie in water so deep as that you cannot reach it to take it up but by diving for it. To encourage yourself in undertaking to do this, reflect that your progress will be from deeper to shallower water, and that at any time you may by bringing your legs under you and standing on the bottom, raise your head far above the water. Then plunge under it with your eyes open, throwing yourself towards the egg, and endeavouring by the action of your hands and feet against the

water

water to get forward till within reach of it. In this attempt you will find, that the water buoys you up againft
your inclination; that it is not fo eafy a thing to fink as you
imagined; that you cannot, but by active force, get down
to the egg. Thus you feel the power of the water to
fupport you, and learn to confide in that power; while
your endeavours to overcome it and to reach the egg, teach
you the manner of acting on the water with your feet and
hands, which action is afterwards ufed in fwimming to fupport your head higher above water, or to go forward
through it.

I would the more earneftly prefs you to the trial of this
method, becaufe, though I think I fatisfyed you that your
body is lighter than water, and that you might float in it a
long time with your mouth free for breathing, if you
would put yourfelf in a proper pofture, and would be ftill
and forbear ftruggling; yet till you have obtained this experimental confidence in the water, I cannot depend on
your having the neceffary prefence of mind to recollect that
pofture and the directions I gave you relating to it. The
furprize may put all out of your mind. For though we
value ourfelves on being reafonable knowing creatures, reafon and knowledge feem on fuch occafions to be of little
ufe to us; and the brutes to whom we allow fcarce a
glimmering of either, appear to have the advantage of us.

I will, however, take this opportunity of repeating thofe
particulars to you, which I mentioned in our laft converfa-

P p p

tion,

tion, as by perufing them at your leifure, you may poffibly imprint them fo in your memory as on occafion to be of fome ufe to you.

1. That though the legs, arms and head, of a human body, being folid parts, are fpecifically fomething heavier than frefh water, yet the trunk, particularly the upper part from its hollownefs, is fo much lighter than water, as that the whole of the body taken together is too light to fink wholly under water, but fome part will remain above, untill the lungs become filled with water, which happens from drawing water into them inftead of air, when a perfon in the fright attempts breathing while the mouth and noftrils are under water.

2. That the legs and arms are fpecifically lighter than falt-water, and will be fupported by it, fo that a human body would not fink in falt-water, though the lungs were filled as above, but from the greater fpecific gravity of the head.

3. That therefore a perfon throwing himfelf on his back in falt-water, and extending his arms, may eafily lie fo as to keep his mouth and noftrils free for breathing; and by a fmall motion of his hands may prevent turning, if he fhould perceive any tendency to it.

4. That in frefh water, if a man throws himfelf on his back, near the furface, he cannot long continue in that fituation but by proper action of his hands on the water. If he ufes no fuch action, the legs and lower part of the body will

will gradually fink till he comes into an upright pofition, in which he will continue fufpended, the hollow of the breaft keeping the head uppermoft.

5. But if in this erect pofition, the head is kept upright above the fhoulders, as when we ftand on the ground, the immerfion will, by the weight of that part of the head that is out of water, reach above the mouth and noftrils, perhaps a little above the eyes, fo that a man cannot long remain fufpended in water with his head in that pofition.

6. The body continuing fufpended as before, and upright, if the head be leaned quite back, fo that the face looks upwards, all the back part of the head being then under water, and its weight confequently in a great meafure fupported by it, the face will remain above water quite free for breathing, will rife an inch higher every infpiration, and fink as much every expiration, but never fo low as that the water may come over the mouth.

7. If therefore a perfon unacquainted with fwimming, and falling accidentally into the water, could have prefence of mind fufficient to avoid ftruggling and plunging, and to let the body take this natural pofition, he might continue long fafe from drowning till perhaps help would come. For as to the cloathes, their additional weight while immerfed is very inconfiderable, the water fupporting it; though when he comes out of the water, he would find them very heavy indeed.

P p p 2

But

But, as I said before, I would not advise you or any one to depend on having this presence of mind on such an occasion, but learn fairly to swim; as I wish all men were taught to do in their youth; they would, on many occurrences, be the safer for having that skill, and on many more the happier, as freer from painful apprehensions of danger, to say nothing of the enjoyment in so delightful and wholesome an exercise. Soldiers particularly should, methinks, all be taught to swim; it might be of frequent use either in surprising an enemy, or saving themselves. And if I had now boys to educate, I should prefer those schools (other things being equal) where an opportunity was afforded for acquiring so advantageous an art, which once learnt is never forgotten.

I am, Sir, &c.

B. F.

L E T-

LETTER LVI.

To Mifs S-------n, at *Wanftead.*

My Dear Friend,   *Sept.* 20, 1761.

IT is, as you obferved in our late converfation, a very general opinion, that *all rivers run into the fea,* or depofite their waters there. 'Tis a kind of audacity to call fuch general opinions in queftion, and may fubject one to cenfure. But we muft hazard fomething in what we think the caufe of truth : And if we propofe our objections modeftly, we fhall, tho' miftaken, deferve a cenfure lefs fevere, than when we are both miftaken and infolent.

That fome rivers run into the fea is beyond a doubt : Such, for inftance, are the *Amazones,* and I think the *Oronoko* and the *Miffifipi.* The proof is, that their waters are frefh quite to the fea, and out to fome diftance from the land. Our queftion is, whether the frefh waters of thofe rivers whofe beds are filled with falt water to a confiderable diftance up from the fea (as the *Thames,* the *Delaware,* and the rivers that communicate with *Chefapeak-bay* in *Virginia)* do ever arrive at the fea ? And as I fufpect

they

they do not, I am now to acquaint you with my reafons, or, if they are not allowed to be reafons, my conceptions at leaft, of this matter.

The common fupply of rivers is from fprings, which draw their origin from rain that has foaked into the earth. The union of a number of fprings forms a river. The waters as they run, expofed to the fun, air and wind, are continually evaporating. Hence in travelling one may often fee where a river runs, by a long blueifh mift over it, tho' we are at fuch a diftance as not to fee the river itfelf. The quantity of this evaporation is greater or lefs, in proportion to the furface expofed by the fame quantity of water to thofe caufes of evaporation. While the river runs in a narrow confined channel in the upper hilly country, only a fmall furface is expofed; a greater as the river widens. Now if a river ends in a lake, as fome do, whereby its waters are fpread fo wide as that the evaporation is equal to the fum of all its fprings, that lake will never over-flow :—And if inftead of ending in a lake, it was drawn into greater length as a river, fo as to expofe a furface equal in the whole to that lake, the evaporation would be equal, and fuch river would end as a canal; when the ignorant might fuppofe, as they actually do in fuch cafes, that the river lofes itfelf by running under ground, whereas in truth it has run up into the air.

Now, many rivers that are open to the fea, widen much before they arrive at it, not merely by the additional waters
they

they receive, but by having their courfe ftopt by the op-
pofing flood tide ; by being turned back twice in twenty-
four hours, and by finding broader beds in the low flat
countries to dilate themfelves in ; hence the evaporation
of the frefh water is proportionably increafed ; fo that in
fome rivers it may equal the fprings of fupply. In fuch
cafes, the falt water comes up the river, and meets the
frefh in that part where, if there were a wall or bank of
earth acrofs from fide to fide, the river would form a lake,
fuller indeed at fome times than at others, according to the
feafons, but whofe evaporation would, one time with another, be equal to its fupply.

When the communication between the two kinds of
water is open, this fuppofed wall of feparation may be conceived as a moveable one, which is not only pufhed fome
miles higher up the river by every flood tide from the fea,
and carried down again as far by every tide of ebb, but
which has even this fpace of vibration removed nearer to the
fea in wet feafons, when the fprings and brooks in the upper country are augmented by the falling rains, fo as to
fwell the river, and farther from the fea in dry feafons.

Within a few miles above and below this moveable line
of feparation, the different waters mix a little, partly by
their motion to and fro, and partly from the greater fpecific gravity of the falt water, which inclines it to run
under the frefh, while the frefh water, being lighter, runs
over the falt.

Caft

Caſt your eye on the map of *North-America*, and obſerve the bay of *Cheſapeak* in *Virginia*, mentioned a-bove ; you will ſee, communicating with it by their mouths, the great rivers *Saſquehanah*, *Potowmack*, *Rappa-hanock*, *York*, and *James*, beſides a number of ſmaller ſtreams, each as big as the *Thames*. It has been pro-poſed by philoſophical writers, that to compute how much water any river diſcharges into the ſea, in a given time, we ſhould meaſure its depth and ſwiftneſs at any part above the tide ; as, for the *Thames*, at *Kingſton* or *Windſor*. But can one imagine, that if all the water of thoſe vaſt rivers went to the ſea, it would not firſt have puſhed the ſalt water out of that narrow-mouthed bay, and filled it with freſh ?—The *Saſquehanah* alone would ſeem to be ſufficient for this, if it were not for the loſs by eva-poration. And yet that bay is ſalt quite up to *Annapolis*.

As to our other ſubject, the different degrees of heat im-bibed from the ſun's rays by cloths of different colours, ſince I cannot find the notes of my experiment to ſend you, I muſt give it as well as I can from memory.

But firſt let me mention an experiment you may eaſily make yourſelf. Walk but a quarter of an hour in your garden when the ſun ſhines, with a part of your dreſs white, and a part black ; then apply your hand to them alternately, and you will find a very great difference in their warmth. The black will be quite hot to the touch, the white ſtill cool.

Another.

Another. Try to fire paper with a burning glafs. If it is white, you will not eafily burn it;—but if you bring the focus to a black fpot, or upon letters, written or printed, the paper will immediately be on fire under the letters.

Thus fullers and dyers find black cloths, of equal thicknefs with white ones, and hung out equally wet, dry in the fun much fooner than the white, being more readily heated by the fun's rays. It is the fame before a fire; the heat of which fooner penetrates black ftockings than white ones, and fo is apt fooner to burn a man's fhins. Alfo beer much fooner warms in a black mug fet before the fire, than in a white one, or in a bright filver tankard.

My experiment was this. I took a number of little fquare pieces of broad cloth from a taylor's pattern card, of various colours. There were black, deep blue, lighter blue, green, purple, red, yellow, white, and other colours, or fhades of colours. I laid them all out upon the fnow in a bright fun-fhiny morning. In a few hours (I cannot now be exact as to the time) the black being warm'd moft by the fun, was funk fo low as to be below the ftroke of the fun's rays; the dark blue almoft as low, the lighter blue not quite fo much as the dark, the other colours lefs as they were lighter; and the quite white remained on the furface of the fnow, not having entered it at all.

Q q q

What

What fignifies philofophy that does not apply to fome ufe ?—May we not learn from hence, that black clothes are not fo fit to wear in a hot funny climate or feafon, as white ones ; becaufe in fuch clothes the body is more heated by the fun when we walk abroad, and are at the fame time heated by the exercife, which double heat is apt to bring on putrid dangerous fevers ? That foldiers and feamen who muft march and labour in the fun, fhould in the *Eaft* or *Weft-Indies* have an uniform of white ? That fummer hats for men or women, fhould be white, as repelling that heat which gives head-achs to many, and to fome the fatal ftroke that the French call the *Coup de Soleil ?* That the ladies Summer hats, however, fhould be lined with black, as not reverberating on their faces thofe rays which are reflected upwards from the earth or water ? That the putting a white cap of paper or linen *within* the crown of a black hat, as fome do, will not keep out the heat, tho' it would if plac'd *without.* That fruit walls being blacked may receive fo much heat from the fun in the day-time, as to continue warm in fome degree thro' the night, and thereby preferve the fruit from frofts, or forward its growth ?—with fundry other particulars of lefs or greater importance, that will occur from time to time to attentive minds ?—

I am,

Yours *affectionately*,

B. FRANKLIN.

LETTER LVII.

Extract of a Letter to Lord *K.* at *Edinburgh*, *June* 2, 1765.

* * * In my paſſage to America I read your excellent work, the *Elements of Criticiſm*, in which I found great entertainment. I only wiſhed you had examined more fully the ſubject of muſick, and demonſtrated that the pleaſure artiſts feel in hearing much of that compoſed in the modern taſte, is not the natural pleaſure ariſing from melody or harmony of ſounds, but of the ſame kind with the pleaſure we feel on ſeeing the ſurprizing feats of tumblers and rope-dancers, who execute difficult things. For my part I take this to be really the caſe, and ſuppoſe it the reaſon why thoſe who are unpractiſed in muſick, and therefore unacquainted with thoſe difficulties, have little or no pleaſure in hearing this muſick. Many pieces of it are mere compoſitions of tricks. I have ſometimes at a concert, attended by a common audience, placed myſelf ſo as to ſee all their faces, and obſerved no ſigns of pleaſure in them during the performance of a great part

that

that was admired by the performers themfelves; while a plain old Scotch tune, which they difdained, and could fcarcely be prevailed on to play, gave manifeft and general delight. Give me leave on this occafion to extend a little the fenfe of your pofition, That "Melody and Harmony are feparately agreable, and in union delightful," and to give it as my opinion that the reafon why the Scotch tunes have lived fo long, and will probably live for ever (if they efcape being ftifled in modern affected ornament) is merely this, that they are really compofitions of melody and harmony united, or rather that their melody is harmony. I mean the fimple tunes fung by a fingle voice. As this will appear parodoxical, I muft explain my meaning. In common acceptation, indeed, only an agreable *fuccefſion* of founds is called *melody*, and only the *co-exiſtence* of agreable founds, *harmony*. But fince the memory is capable of retaining for fome moments a perfect idea of the pitch of a paft found, fo as to compare with it the pitch of a fucceeding found, and judge truly of their agreement or difagreement, there may and does arife from thence a fenfe of harmony between the prefent and paft founds, equally pleafing with that between two prefent founds. Now the conftruction of the old Scotch tunes is this, that almoft every fucceeding emphatical note, is a third, a fifth, an octave, or in fhort fome note that is in concord with the preceding note. Thirds are chiefly ufed, which are very pleafing concords. I ufe the word *emphatical* to diftinguifh

thofe

thofe notes which have a ftrefs laid on them in finging the tune, from the lighter connecting notes, that ferve merely, like grammar articles in common fpeech, to tack the whole together.

That we have a moft perfect idea of a found juft paft, I might appeal to all acquainted with mufick, who know how eafy it is to repeat a found in the fame pitch with one juft heard. In tuning an inftrument, a good ear can as eafily determine that two ftrings are in unifon by founding them feparately, as by founding them together; their difagreement is alfo as eafily, I believe I may fay more eafily and better diftinguifhed when founded feparately; for when founded together, tho' you know by the beating that one is higher than the other, you cannot tell which it is. I have afcribed to memory the ability of comparing the pitch of a prefent tone with that of one paft. But if there fhould be, as poffibly there may be, fomething in the ear fimilar to what we find in the eye, that ability would not be entirely owing to memory. Poffibly the vibrations given to the auditory nerves by a particular found may actually continue fome time after the caufe of thofe vibrations is paft, and the agreement or difagreement of a fubfequent found become by comparifon with them more difcernible. For the impreffion made on the vifual nerves by a luminous object will continue for twenty or thirty feconds. Sitting in a room look earneftly at the middle of a window a little while when the day is bright, and then fhut your eyes;

the

the figure of the window will ftill remain in the eye, and
fo diftinct that you may count the panes. A remarkable
circumftance attending this experiment, is, that the impreffion of forms is better retained than that of colours;
for after the eyes are fhut, when you firft difcern the
image of the window, the panes appear dark, and the
crofs bars of the fafhes, with the window frames and walls,
appear white or bright; but if you ftill add to the darknefs
in the eyes by covering them with your hand, the reverfe
inftantly takes place, the panes appear luminous and the
crofs bars dark. And by removing the hand they are
again reverfed. This I know not how to account for.—
Nor for the following; that after looking long thro' green
fpectacles, the white paper of a book will on firft taking
them off appear to have a blufh of red; and after long
looking thro' red glaffes, a greenifh caft; this feems to
intimate a relation between green and red not yet explained. Farther, when we confider by whom thefe ancient
tunes were compofed, and how they were firft performed,
we fhall fee that fuch harmonical fucceffions of founds
was natural and even neceffary in their conftruction.
They were compofed by the minftrels of thofe days
to be played on the harp accompanied by the voice.
The harp was ftrung with wire, which gives a found
of long continuance, and had no contrivance like that
in the modern harpfichord, by which the found of
the preceding could be ftopt, the moment a fucceeding

note

note began. To avoid actual difcord, it was therefore neceffary that the fucceeding emphatic note fhould be a chord with the preceding, as their founds muft exift at the fame time. Hence arofe that beauty in thofe tunes that has fo long pleafed, and will pleafe for ever, tho' men fcarce know why. That they were originally compofed for the harp, and of the moft fimple kind, I mean a harp without any half notes but thofe in the natural fcale, and with no more than two octaves of ftrings, from C to C, I conjecture from another circumftance, which is, that not one of thofe tunes really ancient, has a fingle artificial half note in it, and that in tunes where it was moft convenient for the voice to ufe the middle notes of the harp, and place the key in F, there the B, which if ufed fhould be a B flat, is always omitted, by paffing over it with a third. The connoiffeurs in modern mufic will fay, I have no tafte, but I cannot help adding, that I believe our anceftors, in hearing a good fong, diftinctly articulated, fung to one of thofe tunes, and accompanied by the harp, felt more real pleafure than is communicated by the generality of modern operas, exclufive of that arifing from the fcenery and dancing. Moft tunes of late compofition, not having this natural harmony united with their melody, have recourfe to the artificial harmony of a bafs, and other accompanying parts*. This fupport, in my opinion, the old tunes do not

need,

* The celebrated *Rouffeau* in his *Dictionaire de Mufique,* printed 1768, appears to have fimilar fentiments of our modern *Harmony,* viz.

" M. Rameau

need, and are rather confused than aided by it. Whoever has heard *James Ofwald* play them on his violoncello, will be lefs inclined to difpute this with me. I have more than once feen tears of pleafure in the eyes of his auditors; and yet, I think, even *his* playing thofe tunes would pleafe more, if he gave them lefs modern ornament.

I am, &c.

B. F.

" M. Rameau pretend que les deffus d'une certaine fimplicité fuggèrent naturellement leur baffe, & qu'un homme ayant l'oreille jufte & non exercée, entonnera naturellement cette baffe. C'eft-là un préjugé de muficien, démenti par toute expérience. Non feulement celui qui n'aura jamais entendu ni baffe ni *harmonie*, ne trouvera, de lui-même, ni cette *harmonie* ni cette baffe; mais elles lui déplairont fi on les lui fait entendre, & il aimera beaucoup mieux le fimple uniffon.

Quand on fonge que, de tous les peuples de la terre, qui tous ont une mufique & un chant, les Européens font les feuls qui aient une *harmonie* des accords, & qui trouvent ce mélange agréable; quand on fonge que le monde a duré tant de fiècles, fans que, de toutes les nations qui ont cultivé les beaux arts, aucune ait connu cette *harmonie*; qu'aucun animal, qu'aucun oifeau, qu'aucun être dans la nature ne produit d'autre accord que l'uniffon, ni d'autre mufique que la mélodie; que les langues orientales, fi fonores, fi muficales; que les oreilles Grecques, fi délicates, fi fenfibles, exercées avec tant d'art, n'ont jamais guidé ces peuples voluptueux & paffionnés vers notre *harmonie*; que, fans elle, leur mufique avoit des effets fi prodigieux: qu'avec elle la nôtre en a de fi foibles; qu'enfin il étoit réfervé à des peuples du Nord, dont les organes durs & groffiers font plus touchés de l'éclat & du bruit des voix, que de la douceur des accens, & de la mélodie des inflexions, de faire cette grande découverte, & de la donner pour principe à toutes les règles de l'art; quand, dis-je, on fait attention à tout cela, il eft bien difficile de ne pas foupçonner que toute notre *harmonie* n'eft qu'une invention gothique & barbare, dont nous ne nous fuffions jamais avifés, fi nous euffions été plus fenfibles aux véritables beautés de l'art, & à la mufique vraiment naturelle."

LETTER LVI.

To Mr. P. F. Newport, *New England.*

Dear Brother,

* * * * " I like your ballad, and think it well adapted for your purpose of difcountenancing expenfive foppery, and encouraging induftry and frugality. If you can get it generally fung in your country, it may probably have a good deal of the effect you hope and expect from it. But as you aimed at making it general, I wonder you chofe fo uncommon a meafure in poetry, that none of the tunes in common ufe will fuit it. Had you fitted it to an old one, well known, it muft have fpread much fafter than I doubt it will do from the beft new tune we can get compos'd for it. I think too, that if you had given it to fome country girl in the heart of the *Maffachufets,* who has never heard any other than pfalm tunes, or *Chevy Chace,* the *Children in the Wood,* the *Spanifh Lady,* and fuch old fimple ditties, but has naturally a good ear, fhe might more probably have made a pleafing popular tune for you, than any of our mafters here, and more proper for your purpofe, which

R r r would

would beſt be anſwered, if every word could as it is ſung be underſtood by all that hear it, and if the emphaſis you intend for particular words could be given by the ſinger as well as by the reader; much of the force and impreſſion of the ſong depending on thoſe circumſtances. I will however get it as well done for you as I can.

Do not imagine that I mean to depreciate the ſkill of our compoſers of muſic here; they are admirable at pleaſing *practiſed* ears, and know how to delight *one another*; but, in compoſing for ſongs, the reigning taſte ſeems to be quite out of nature, or rather the reverſe of nature, and yet like a torrent, hurries them all away with it; one or two perhaps only excepted.

You, in the ſpirit of ſome ancient legiſlators, would influence the manners of your country by the united powers of poetry and muſic. By what I can learn of *their* ſongs, the muſic was ſimple, conformed itſelf to the uſual pronunciation of words, as to meaſure, cadence or emphaſis, *&c.* never difguiſed and confounded the language by making a long ſyllable ſhort, or a ſhort one long when ſung; their ſinging was only a more pleaſing, becauſe a melodious manner of ſpeaking; it was capable of all the graces of proſe oratory, while it added the pleaſure of harmony. A modern ſong, on the contrary, neglects all the proprieties and beauties of common ſpeech, and in their place introduces its *defects* and *abſurdities* as ſo many graces. I am afraid you will hardly take my word for this,

and

and therefore I muft endeabour to fupport it by proof.
Here is the firft fong I lay my hand on.   It happens to be
a compofition of one of our greateft mafters, the ever famous *Handel.*   It is not one of his juvenile performances,
before his tafte could be improved and formed: It appeared when his reputation was at the higheft, is greatly admired by all his admirers, and is really excellent in its kind.
It is called, *The additional* FAVOURITE *Song in* Judas
Maccabeus.   Now I reckon among the defects and im-
proprieties of common fpeech, the following, viz.

1. *Wrong placing the accent or emphafis,* by laying it on
words of no importance, or on wrong fyllables.

2. *Drawling ;* or extending the found of words or fyllables beyond their natural length.

3. *Stuttering ;* or making many fyllables of one.

4. *Unintelligiblenefs ;* the refult of the three foregoing
united.

5. *Tautology ;* and

6. *Screaming,* without caufe

For the *wrong placing of the accent, or emphafis,* fee it
on the word *their* inftead of being on the word *vain.*

with   *their*   vain   My - fte - rious   Art

And on the word *from*, and the wrong fyllable *like*.

God-*like* Wifdom *from* a — bove.

For the *Drawling*, fee the laft fyllable of the word *wounded*.

Nor can heal the wound*ed* Heart

And in the fyllable *wif*, and the word *from*, and fyllable *bove*

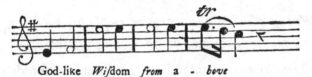

God-like *Wif*dom *from* a - *bove*

For the *Stuttering*, fee the words *ne'er relieve*, in

Ma - gick Charms can *ne er re - lieve* you

Here are four fyllables made of one, and eight of three
but this is moderate. I have feen in another fong that I
cannot now find, feventeen fyllables made of three, and fixteen of one; the latter I remember was the word *charms*;
viz. *Cha, a, a, a, a, a, a, a, a, a, a, a, a, a, a, arms.* Stam-
mering with a witnefs!

For the *Unintelligiblenefs*; give this whole fong to any
taught

taught finger, and let her fing it to any company that have never heard it; you fhall find they will not under-ftand three words in ten. It is therefore that at the oratorio's and operas one fees with books in their hands all thofe who defire to underftand what they hear fung by even our beft performers.

For the *Tautology*; you have, *with their vain myfterious art,* twice repeated; *Magic charms can ne'er relieve you,* three times. *Nor can heal the wounded heart,* three times. *Godlike wifdom from above,* twice; and, *this alone can ne'er deceive you,* two or three times. But this is reafonable when compared with *the Monfter Polypheme, the Monfter Polypheme,* a hundred times over and over, in his admired *Acis and Galatea.*

As to the *fcreaming*; perhaps I cannot find a fair inftance in this fong; but whoever has frequented our operas will remember many. And yet here methinks the words *no* and *e'er,* when fung to thefe notes, have a little of the air of *fcreaming,* and would actually be fcream'd by fome fingers.

No magic charms can *e'er* re—lieve you.

I fend you inclofed the fong with its mufic at length. Read the words without the repetitions. Obferve how

few

few they are, and what a fhower of notes attend them: You will then perhaps be inclined to think with me, that though the words might be the principal part of an ancient fong, they are of fmall importance in a modern one; they are in fhort only *a pretence for finging.*

> *I am, as ever,*
> *Your affectionate brother,*
> B. F.

P. S. I might have mentioned *Inarticulation* among the defects in common fpeech that are affumed as beauties in modern finging. But as that feems more the fault of the finger than of the compofer, I omitted it in what related merely to the compofition. The fine finger in the prefent mode, ftifles all the hard confonants, and polifhes away all the rougher parts of words that ferve to diftinguifh them one from another; fo that you hear nothing but an admirable pipe, and underftand no more of the fong, than you would from its tune played on any other inftrument. If ever it was the ambition of muficians to make inftruments that fhould imitate the human voice, that ambition feems now reverfed, the voice aiming to be like an inftrument. Thus wigs were firft made to imitate a good natural head of hair;—but when they became fafhionable, though in unnatural forms, we have feen natural hair dreffed to look like wigs.

Of

LETTER LIX.

Of LIGHTNING, *and the Method (now ufed in* America) *of fecuring Buildings and Perfons from its mifchievous Effects.*

EXPERIMENTS made in electricity firft gave phi-lofophers a fufpicion that the matter of lightning was the fame with the electric matter. Experiments after-wards made on lightning obtained from the clouds by pointed rods, received into bottles, and fubjected to every trial, have fince proved this fufpicion to be perfectly well founded; and that whatever properties we find in electri-city, are alfo the properties of lightning.

This matter of lightning, or of electricity, is an extream fubtile fluid, penetrating other bodies, and fubfifting in them, equally diffufed.

When by any operation of art or nature, there happens to be a greater proportion of this fluid in one body than in another, the body which has moft, will communicate to that which has leaft, till the proportion becomes equal; provided the diftance between them be not too great; or, if it is too great, till there be proper conductors to con-vey it from one to the other.

If

If the communication be through the air without any conductor, a bright light is seen between the bodies, and a sound is heard. In our small experiments we call this light and sound the electric spark and snap; but in the great operations of nature, the light is what we call· *lightning*, and the sound (produced at the same time, tho' generally arriving later at our ears than the light does to our eyes) is, with its echoes, called *thunder*.

If the communication of this fluid is by a conductor, it may be without either light or sound, the subtle fluid passing in the substance of the conductor.

If the conductor be good and of sufficient bigness, the fluid passes through it without hurting it. If otherwise, it is damaged or destroyed.

All metals, and water, are good conductors.—Other bodies may become conductors by having some quantity of water in them, as wood, and other materials used in building, but not having much water in them, they are not good conductors, and therefore are often damaged in the operation.

Glass, wax, silk, wool, hair, feathers, and even wood, perfectly dry are non-conductors: that is, they resist instead of facilitating the passage of this suble fluid.

When this fluid has an opportunity of passing through two conductors, one good, and sufficient, as of metal, the other not so good, it passes in the best, and will follow it in any direction.

The

The diftance at which a body charged with this fluid will difcharge itfelf fuddenly, ftriking through the air into another body that is not charged, or not fo highly charg'd, is different according to the quantity of the fluid, the dimenfions and form of the bodies themfelves, and the ftate of the air between them —This diftance, whatever it happens to be between any two bodies, is called their *ftriking diftance,* as till they come within that diftance of each o-ther, no ftroke will be made.

The clouds have often more of this fluid in proportion than the earth; in which cafe as foon as they come near enough (that is, within the ftriking diftance) or meet with a conductor, the fluid quits them and ftrikes into the earth. A cloud fully charged with this fluid, if fo high as to be beyond the ftriking diftance from the earth, paffes quietly without making noife or giving light; unlefs it meets with other clouds that have lefs.

Tall trees, and lofty buildings, as the towers and fpires of churches, become fometimes conductors between the clouds and the earth; but not being good ones, that is, not conveying the fluid freely, they are often damaged.

Buildings that have their roofs covered with lead, or other metal, and fpouts of metal continued from the roof into the ground to carry off the water, are never hurt by lightning, as whenever it falls on fuch a building, it paffes in the metals and not in the walls.

S f f   When

When other buildings happen to be within the ſtriking diſtance from ſuch clouds, the fluid paſſes in the walls whether of wood, brick or ſtone, quitting the walls only when it can find better conductors near them, as metal rods, bolts, and hinges of windows or doors, gilding on wainſcot, or frames of pictures; the ſilvering on the backs of look-ing-glaſſes; the wires for bells; and the bodies of animals, as containing watry fluids. And in paſſing thro' the houſe it follows the direction of theſe conductors, taking as many in it's way as can aſſiſt it in its paſſage, whether in a ſtrait or crooked line, leaping from one to the other, if not far diſtant from each other, only rending the wall in the ſpaces where theſe partial good conductors are too diſtant from each other.

An iron rod being placed on the outſide of a building, from the higheſt part continued down into the moiſt earth, in any direction ſtrait or crooked, following the form of the roof or other parts of the building, will receive the light-ning at its upper end, attracting it ſo as to prevent its ſtrik-ing any other part; and, affording it a good conveyance into the earth, will prevent its damaging any part of the building.

A ſmall quantity of metal is found able to conduct a great quantity of this fluid. A wire no bigger than a gooſe quill, has been known to conduct (with ſafety to the building as far as the wire was continued) a quantity of lightning that did prodigious damage both above and

below

below it; and probably larger rods are not neceffary, tho' it is common in America, to make them of half an inch, fome of three quarters, or an inch diameter.

The rod may be faftened to the wall, chimney, &c. with ftaples of iron.—The lightning will not leave the rod (a good conductor) to pafs into the wall (a bad conductor), through thofe ftaples.—It would rather, if any were in the wall, pafs out of it into the rod to get more readily by that conductor into the earth.

If the building be very large and extenfive, two or more rods may be placed at different parts, for greater fecurity.

Small ragged parts of clouds fufpended in the air between the great body of clouds and the earth (like leaf gold in electrical experiments), often ferve as partial conductors for the lightning, which proceeds from one of them to another, and by their help comes within the ftriking diftance to the earth or a building. It therefore ftrikes through thofe conductors a building that would otherwife be out of the ftriking diftance.

Long fharp points communicating with the earth, and prefented to fuch parts of clouds, drawing filently from them the fluid they are charged with, they are then attracted to the cloud, and may leave the diftance fo great as to be beyond the reach of ftriking.

It is therefore that we elevate the upper end of the rod fix or eight feet above the higheft part of the building, ta-

pering

pering it gradually to a fine sharp point, which is gilt to prevent its rusting.

Thus the pointed rod either prevents a stroke from the cloud, or, if a stroke is made, conducts it to the earth with safety to the building.

The lower end of the rod should enter the earth so deep as to come at the moist part, perhaps two or three feet; and if bent when under the surface so as to go in a horizontal line six or eight feet from the wall, and then bent again downwards three or four feet, it will prevent damage to any of the stones of the foundation.

A person apprehensive of danger from lightning, happening during the time of thunder to be in a house not so secured, will do well to avoid sitting near the chimney, near a looking glass, or any gilt pictures or wainscot; the safest place is in the middle of the room, (so it be not under a metal lustre suspended by a chain) sitting in one chair and laying the feet up in another. It is still safer to bring two or three mattrasses or beds into the middle of the room, and folding them up double, place the chair upon them; for they not being so good conductors as the walls, the lightning will not chuse an interrupted course through the air of the room and the bedding, when it can go thro' a continued better conductor the wall. But where it can be had, a hamock or swinging bed, suspended by silk cords equally distant from the walls on every side, and from the cieling and floor above and below, affords the safest

situation

fituation a perfon can have in any room whatever; and what indeed may be deemed quite free from danger of any ftroke by lightning.

Paris, Sept. 1767.   B. F.

LETTER LX.

Extract of a Letter from J. W. *Efq*; *Profeffor of Natural Philofophy at* Cambridge, *in* New England. *Jan.* 6, 1768.

" * * * I have read in the Philofophical Tranfactions
" the account of the effects of lightning on St. Bride's
" fteeple. 'Tis amazing to me, that after the full demon-
" ftration you had given, of the identity of lightning and of
" electricity, and the power of metalline conductors, they
" fhould ever think of repairing that fteeple without fuch
" conductors.   How aftonifhing is the force of prejudice
" even in an age of fo much knowledge and free en-
" quiry!"

ANSWER

ANSWER to the above.

* * * It is perhaps not fo extraordinary that unlearned men, fuch as commonly compofe our church veftries, fhould not yet be acquainted with, and fenfible of the benefits of metal conductors in averting the ftroke of lightning, and preferving our houfes from its violent effects, or that they fhould be ftill prejudiced againft the ufe of fuch conductors, when we fee how long even philofophers, men of extenfive fcience and great ingenuity, can hold out againft the evidence of new knowledge, that does not fquare with their preconceptions; and how long men can retain a practice that is conformable to their prejudices, and expect a benefit from fuch practice, though conftant experience fhows its inutility. A late piece of the Abbé *Nollet*, printed laft year in the memoirs of the French Academy of fciences, affords ftrong inftances of this: For though the very relations he gives of the effects of lightning in feveral churches and other buildings, fhow clearly that it was conducted from one part to another by wires, gildings, and other pieces of metal that were *within*, or connected with the building, yet in the fame paper he objects to the providing metalline conductors *without* the building, as ufelefs or dangerous*. He cautions people not to ring the church

bells

---

* Notre curiofité pourroit peut-être s'applaudir des recherches qu'elle nous a fait faire fur la nature du tonnerre, & fur la mécanifme de fes principaux effets, mais ce néft point ce qu il y a de plus important; il vaudroit bien mieux

que

bells during a thunder-ftorm, left the lightning, in its way
to the earth, fhould be conducted down to them by the
bell ropes*, which are but bad conductors; and yet is a-
gainft fixing metal rods on the outfide of the fteeple, which
are known to be much better conductors, and which it
would certainly chufe to pafs in, rather than in dry hemp.
And though for a thoufand years paft bells have been fo-
lemnly confecrated by the Romifh church†, in expectation
that

que nous puiffions trouver quelque moyen de nous en garantir : on ly a penfé;
on s'eft même flatté d'avoir fait cette grande decouverte; mais malheureufe-
ment douze années d'épreuves & un peu de réflexion, nous apprennent qu'il
ne faut pas compter fur les promeffes qu'on nous a faites. Je l'ai dit, il y a
long temps, and avec regret, toutes ces pointes de fer qu'on dreffe en l'air, foit
comme *électrofcopes*, foit comme préfervatifs, *** font plus propre à nous attirer
le feu du tonnerre qu'à nous en préferver; *** & je perfifte à dire que le projet
d'épuifer une nuée orageufe du feu dont elle eft chargée, n'eft pas celui d'un
phyficien.—***. *Memoire fur les Effets du Tonnerre.*

* Les cloches, en vertu de leur bénédiction, doivent écarter les orages &
nous preferver des coups de foudre; mais l'eglife permet à la prudence humaine
le choix des momens où il convient d'ufer de ce préfervatif. Je ne fais fi le
fon, confidéré phyfiquement, eft capable ou non de faire crever une nuée & de
caufer l'épanchement de fon feu vers les objets terreftres, mais il eft certain &
prouvé par l'expérience, que la tonnerre peut tomber fur un clocher, foit que
l'on y fonne ou que l'on n'y fonne point; & fi cela arrive dans le premier cas,
les fonneurs font en grand danger, parcequ'ils tiennent des cordes par lefquelles
la commotion de la foudre peut fe communiquer jufqu'à eux : il eft donc plus
fage de laiffer les cloches en repos quand l'orage eft arrivé au-deffus de
l'églife. Ibid.

† Suivant le rituel de Paris, lorfqu'on benit des cloches, on recite les orai-
fons fuivantes :

Benedic

that the found of fuch bleffed bells would drive away thofe storms, and fecure our buildings from the ftroke of lightning; and during fo long a period, it has not been found by experience, that places within the reach of fuch bleffed found, are fafer than others where it is never heard; but that on the contrary, the lightning feems to ftrike fteeples of choice, and that at the very time the bells are ringing†; yet ftill they continue to blefs the new bells, and jangle the old ones whenever it thunders.—One would think it was now time to try fome other trick;—and ours is recommended (whatever this able philofopher may have been told to the contrary) by more than twelve years experience, wherein, among the great number of houfes furnifhed with iron rods in North America, not one fo guarded has been materially hurt with lightning, and feve-

Benedic, Domine . . . . . quotiefcumque fonueri!, procul recedat virtus infidiantium, umbra phantafmatis, incurfio turbinum, percuffio fulminum, læfio tonitruum, calamitas tempeftatum, omnifque fpiritus procellarum, &c.

Deus, qui per beatum Moïfen, &c. . . . . . . procul pellentur infidiæ inimici, fragor grandinum, procella turbinum, impetus tempeftatum, temperentur infefta tonitrua, &c.

Omnipotens fempiterne Deus, &c. . . . . . . ut ante fonitum ejus effugentur ignita jacula inimici, percuffio fulminum, impetus lapidum, læfio tempeftatum, &c.

† En 1718. M. Deflandes fit favoir à l'Academie Royale des fciences, que la nuit du 14 ou 15 d'Avril de la même année, le tonnerre étoit tombé fur vingtquatre églifes, depuis Landernau jufqu'à Saint-Pol-de-Léon en Bretagne; que ces églifes étoient précifément celles où l'on fonnoit, & que la foudre avoit épargné celles ou l'on ne fonnoit pas: que dans celle de Gouifno.., qui fut entièrement ruinée, le tonnerre tua deux perfonnes de quatre qui fonnoient, &c. *Hift. de l'Ac. R. des Sci.* 1719.

ral

ral have been evidently preferved by their means; while a number of houfes, churches, barns, fhips, &c. in different places, unprovided with rods, have been ftruck and greatly damaged, demolifhed or burnt. Probably the veftries of our Englifh churches are not generally well acquainted with thefe facts; otherwife, fince as good proteftants they have no faith in the blefling of bells, they would be lefs excufable in not providing this other fecurity for their refpective churches, and for the good people that may happen to be affembled in them during a tempeft, efpecially as thofe buildings, from their greater height, are more expofed to the ftroke of lightning than our common dwellings.

I have nothing new in the philofophical way to communicate to you, except what follows. When I was laft year in *Germany*, I met with a fingular kind of glafs, being a tube about eight inches long, half an inch in diameter, with a hollow ball of near an inch diameter at one end, and one of an inch and half at the other, hermetically fealed, and half filled with water.—If one end is held in hand, and the other a little elevated above the level, a conftant fucceflion of large bubbles proceeds from the end in the hand to the other end, making an appearance that puzzled me much, 'till I found that the fpace not filled with water was alfo free from air, and either filled with a fubtile invifible vapour continually rifing from the water, and extreamly rarifiable by the leaft heat at

T t t

one

one end, and condenfable again by the leaft coolnefs at the other; or it is the very fluid of fire itfelf, which parting from the hand pervades the glafs, and by its expanfive force depreffes the water till it can pafs between it and the glafs, and efcape to the other end, where it gets thro' the glafs again into the air. I am rather inclined to the firft opinion, but doubtful between the two. An ingenious artift here, Mr. *Nairne*, mathematical inftrument-maker, has made a number of them from mine, and improved them, for his are much more fenfible than thofe I brought from Germany.—I bor'd a very fmall hole through the wainfcot in the feat of my window, through which a little cold air conftantly entered, while the air in the room was kept warmer by fires daily made in it, being winter time. I plac'd one of his glaffes, with the elevated end againft this hole; and the bubbles from the other end, which was in a warmer; fituation, were continually paffing day and night, to the no fmall furprize of even philofophical fpectators. Each bubble difcharged, is larger than that from which it proceeds, and yet that is not diminifhed; and by adding itfelf to the bubble at the other end, that bubble is not increafed, which feems very paradoxical.—When the balls at each end are made large, and the connecting tube very fmall and bent at right angles, fo that the balls, inftead of being at the ends, are brought on the fide of the tube, and the tube is held fo as that the balls are above it, the water will be depreffed in that which is held in the hand,

hand, and rife in the other as a jet or fountain; when it is all in the other, it begins to boil, as it were, by the vapour paffing up through it; and the inftant it begins to boil, a fudden coldnefs is felt in the ball held; a curious experiment, this, firft obferved and fhewn me by Mr. *Nairne.* There is fomething in it fimilar to the old obfervation, I think mentioned by *Ariftotle,* that the bottom of a boiling pot is not warm; and perhaps it may help to explain that fact;—if indeed it be a fact.——When the water ftands at an equal height in both thefe balls, and all at reft; if you wet one of the balls by means of a feather dipt in fpirit, though that fpirit is of the fame temperament as to heat and cold, with the water in the glaffes, yet the cold occafioned by the evaporation of the fpirit from the wetted ball, will fo condenfe the vapour over the water contained in that ball, as that the water of the other ball will be preffed up into it, followed by a fucceffion of bubbles, 'till the fpirit is all dried away. Perhaps the obfervations on thefe little inftruments may fuggeft and be applied to fome beneficial ufes. It has been thought that water reduced to vapour by heat, was rarified only fourteen thoufand times, and on this principle our engines for raifing water by fire are faid to be conftructed: But if the vapour fo much rarified from water, is capable of being itfelf ftill farther rarified to a boundlefs degree by the application of heat to the veffels or parts of veffels containing the vapour (as at firft it is applied to thofe containing the water) perhaps a

T t t 2

muc

a much greater power may be obtained, with little additional expence. Poſſibly too, the power of eaſily moving
water from one end to the other of a moveable beam (ſuſpended in the middle like a ſcale beam) by a ſmall degree
of heat, may be applied advantageouſly to ſome other mechanical purpoſes. * * *

I am, &c.   B. F.

L E T T E R LXI.

To Sir *John Pringle*, Bart.

S I R,   *Craven-ſtreet, May* 10, 1768.

YOU may remember that when we were travelling
together in *Holland*, you remarked that the track-
ſchuyt in one of the ſtages went ſlower than uſual, and
enquired of the boatman, what might be the reaſon; who
anſwered, that it had been a dry ſeaſon, and the water in
the canal was low.   On being again aſked if it was ſo low
as that the boat touch'd the muddy bottom; he ſaid, no, not
ſo low as that, but ſo low as to make it harder for the
horſe to draw the boat.   We neither of us at firſt could
conceive that if there was water enough for the boat to
ſwim

ſwim clear of the bottom, its being deeper would make any
difference; but as the man affirmed it ſeriouſly as a thing
well known among them; and as the punctuality required
in their ſtages, was likely to make ſuch difference, if any
there were, more readily obſerved by them, than by other
watermen who did not paſs ſo regularly and conſtantly
backwards and forwards in the ſame track; I began to apprehend there might be ſomething in it, and attempted to
account for it from this conſideration, that the boat in proceeding along the canal, muſt in every boat's length of her
courſe, move out of her way a body of water, equal in
bulk to the room her bottom took up in the water; that
the water ſo moved, muſt paſs on each ſide of her and under her bottom to get behind her; that if the paſſage under
her bottom was ſtraitened by the ſhallows, more of that
water muſt paſs by her ſides, and with a ſwifter motion,
which would retard her, as moving the contrary way; or
that the water becoming lower behind the boat than before, ſhe was preſſed back by the weight of its difference
in height, and her motion retarded by having that weight
conſtantly to overcome.   But as it is often loſt time to attempt accounting for uncertain facts, I determined to make
an experiment of this when I ſhould have convenient time
and opportunity.

   After our return to *England*, as often as I happened to be
on the *Thames*, I enquired of our watermen whether they
were ſenſible of any difference in rowing over ſhallow or

deep

water. I found them all agreeing in the fact, that there was a very great difference, but they differed widely in expreffing the quantity of the difference; fome fuppofing it was equal to a mile in fix, others to a mile in three, &c. As I did not recollect to have met with any mention of this matter in our philofophical books, and conceiving that if the difference fhould really be great, it might be an object of confideration in the many projects now on foot for digging new navigable canals in this ifland, I lately put my defign of making the experiment in execution, in the following manner.

I provided a trough of plained boards fourteen feet long, fix inches wide and fix inches deep, in the clear, filled with water within half an inch of the edge, to reprefent a canal. I had a loofe board of nearly the fame length and breadth, that being put into the water might be funk to any depth, and fixed by little wedges where I would chufe to have it ftay, in order to make different depths of water, leaving the furface at the fame height with regard to the fides of the trough. I had a little boat in form of a lighter or boat of burthen, fix inches long, two inches and a quarter wide, and one inch and a quarter deep. When fwimming, it drew one inch water. To give motion to the boat, I fixed one end of a long filk thread to its bow, juft even with the water's edge, the other end paffed over a well-made brafs pully, of about an inch diameter, turning freely on a fmall axis; and a fhilling was the weight. Then placing the boat

at

at one end of the trough, the weight would draw it through the water to the other.

Not having a watch that shows seconds, in order to measure the time taken up by the boat in passing from end to end, I counted as fast as I could count to ten repeatedly, keeping an account of the number of tens on my fingers. And as much as possible to correct any little inequalities in my counting, I repeated the experiment a number of times at each depth of water, that I might take the medium.— And the following are the results.

Water 1½ inches deep.	2 inches.	4½ inches.
1st exp. - 100	- - - 94	- - - 79
2 - - - 104	- - - 93	- - - 78
3 - - - 104	- - - 91	- - - 77
4 - - - 106	- - - 87	- - - 79
5 - - - 100	- - - 88	- - - 79
6 - - - 99	- - - 86	- - - 80
7 - - - 100	- - - 90	- - - 79
8 - - - 100	- - - 88	- - - 81
813	717	632

Medium 101   Medium 89   Medium 79

I made many other experiments, but the above are those in which I was most exact; and they serve sufficiently to show that the difference is considerable. Between the deepest and shallowest it appears to be somewhat more

than

than one fifth. So that fuppofing large canals and boats and depths of water to bear the fame proportions, and that four men or horfes would draw a boat in deep water four leagues in four hours, it would require five to draw the fame boat in the fame time as far in fhallow water; or four would require five hours.

Whether this difference is of confequence enough to juftify a greater expence in deepening canals, is a matter of calculation, which our ingenious engineers in that way will readily determine.

I am, &c. B. F.

T H E E N D.

INDEX.

A

ÆTHER, what, 277.

Accent or *Emphafis*, wrong placing of it, one of the faults in modern fongs, 475.

Air, its particles, their properties, 41, 182, & *feq*. its currents over the globe, 47, 48.

—It refifts the motion of the electric fluid, 80, and confines it to bodies, 81, 174.

—Its effects in electric experiments, 96, 97.

— Its friction againft trees, &c. confidered, 112, 176.

Air, and water mutually attract each other, 182. diffolves water, 183, and, when dry, oil, 185. why fuffocating when impregnated with oil, 185.

— how water is fupported in it, 186. Objections to this opinion, 247.

— of hurricanes cold, and why, 188.

Air, in rooms, electrified pofitively and negatively by Mr. *Canton*, 409.

— effects of heat upon it, 285.

— has its fhare of electricity, 386, experiments to prove it, *ibid*.

—Its electricity fometimes denfer above than below, 388, 410.

— rare, no bad conductor of found, 437.

Air-Thermometer, electrical, experiments with it, 390, & *feq*.

Amber, electric experiments on it, 425.

America, why marriage is more early there, and more general, 198.

Analyfis of the Leiden phial, 26, 27.

Animal-Heat, whence it arifes, 346.

Animalcules, fuppofed to caufe the luminous appearance of fea water, 274, & *feq*.

Armies, whence the beft means of fupporting them, 338.

Armonica, the, a new mufical inftrument defcribed, 428, & *feq*. Manner of playing on it, 433.

Atmofphere, electric, its properties, 155.

———— fometimes denfer above than below, 187.

Aurora-

I N D E X.

Colonies,

INDEX.

INDEX.

I N D E X.

Fire,

INDEX.

INDEX.

Glass, the manner of its operation in producing electricity, explain'd, 76, & *seq.*

—— its elasticity, to what possibly owing, 78.

—— thick, resists the change in the electricity of its different sides, 82.

—— rod of it, will not conduct a shock, *ibid.*

—— rubbed, is positively electrified, sulphur negatively, reasons for thinking so, 104, 105.

—— globes of it, some will not produce electricity, 319, what best, *ibid.*

—— an error relating to its pores, acknowledg'd, 321.

Glass, a curious one describ'd, 489, & *seq.*

Glasses, a new musical instrument compos'd of them, described, 428, & *seq.*

Gold Leaf, its suspension explain'd, 70, & *seq.*

Government, free, can only be destroyed by corrupt manners, 335.

Greece, its superiority over *Persia*, whence, 335.

Green and *red*, a relation between those colours, 470.

Gunpowder, dry, how to be fired by electricity, 92.

Gunpowder, magazines of, how to secure them from lightning,

441, & *seq.* New proposal for keeping gunpowder dry, 442.

H

Habits, and manners, their efficacy in increasing mankind, 331, 332.

Harmony, and melody, what, 469.

Harp, ancient Scotch tunes composed to be sung with that instrument, 470. The effect it had on their composition, *ibid.*

Hats, ladies summer, of what colour best, 466.

Heat, the pain it occasions, how produced, 345.

—— in an animal, how produced, 346, 347.

—— in fermentation, its degree, 347.

—— great, at Philadelphia, June 1750, its circumstances, 365, 366.

—— produced in bodies by electricity, and by lightning, 392, 411, 413, 419.

Hole, struck through pasteboard, reason of the bur round it, 123

Horse-Race, electrical, 388.

Hudson's-River, winds there, 255, 259, 278.

Hunters,

INDEX.

X x x   *Lightning*,

North-

I N D E X.

ments on the clouds, 112.
its ufe as a conductor of
lightning, 482.

Rome, effect of its manners confidered, 335.

Rooms, warm, their advantages,
261, *& feq.* Do not give
people colds, 310, *& feq.*

Rouffeau, his opinion of tunes
in parts, 472.

S

Salt, Rock, form'd originally
from the fea, and in what
manner, 379, 380.

Salt-Water, how to quench
thirft with it, 459.

Scotch Tunes, the pleafure they
give explain'd, 468.

Sea, the fource of lightning,
objections to that opinion,
170. Reafons on which it
was founded, 174, 175.
——— its luminous appearance
accounted for, 273, *& feq*
——— not falt from the diffolution of rock-falt, but frefher
now than it was originally,
379.
——— has formerly covered the
mountains. *ibid.*

Sea-Water, foon lofes its luminous quality, 111, 176.

Slaves, not profitable labourers,
200.
——— diminifh and vitiate the
free people, 202.

Small-Pox, lofs by it in *Bofton*,
193, 194.

Smell of Electricity, produc'd by
its action on fomething in the
air, 84.

Snow, blown about by the wind,
gave electricity, tho' the air
otherwife clear, 434.

Songs ancient, gave more pleafure than modern ones, why,
471, 474. Modern, com-
pos'd of all the defects and
abfurdities of common fpeech,
ibid. Inftances of it, 475,
& feq.

Sound, obfervations on it, 435,
under water ftrong at a diftance, *ibid.* Queries concerning it, 436, 437. A fubtle
fluid fufpected as the conductor of it, *ibid:*

Sounds, juft paft, we have a perfect idea of their pitch, 468,
469.

Spain, what has thinned its peo-
-ple, 205.

Specifical Weight, what, 286.

Spheres electric, commodious ones,
11.

Spider, counterfeit, defcribed, 10.

Spirit, inflammable, linen cloths
wet with it, and applied to
painful imflammations of the
body, give great coolnefs and
eafe, 368.

Spirits, fired without heating
them, 51.

Spots,

INDEX.

Spots, in the Sun, how formed, 266.

Stars, shooting, what, 236.

St. Bride's Church, observation of the stroke of lightning on it, 440, 441.

Stuttering, one of the affected beauties of modern songs, 475, & *seq.*

Sun, why not wasted by expence of light, 266.

Surface, whether increase of it can account for the rise of vapours, 248, 251, 259.

Surfaces, the opposite ones of glass, their different state when electris'd, 25, 75.

Sulphur Globe, its electricity different and opposite to that of the glass globe, 103.

Sweat, the necessity of keeping it up, in those that labour in the sun in hot climates, 366. The manner of doing it, *ibid.*

Swiftness of electric motion explain'd, 282, 283.

Swimming, how to be learnt, 463, & *seq.* How a person unacquainted with it may avoid sinking, 466, 467. A delightful and wholesome exercise, 468. Its advantage to soldiers, *ibid.*

T

Tautology, one of the beauties of modern songs, 475, 477.

Thermometer, not cooled by blowing on it when dry, 368.

—— electrical, described, 389. Experiments made with it, 390, & *seq.*

Thirst, how it may be quenched with salt water, 459, 460.

Thunder-gusts, what, 39.

—— a new hypothesis for explaining them, *ib.* & *seq.*

Tides, in rivers, their motion explain'd, 450, & *seq.* A tide is a wave, high and low water different parts of that wave, *ibid.* Not much more water in a river at high water than at low water 454, 455, two, three, or more high waters, and as many low waters existing in the same river at the same time, 452, 457.

Tourmalin, some experiments on it, 375, & *seq.*

Towns, in England, reason of their growth, 334.

Trees, dangerous to take shelter under them, 50. Why cool in the sun, 367.

—— the shivering of them by lightning, explained, 415.

Triangles equilateral, form'd by particles of air, 42.

Tube, of glass, lin'd with a non-electric, experiment with it, 79.

—— may be made by rubbing to act like the *Leiden* bottle, 80.

Tube,

INDEX.

INDEX.

Printed in the United States
By Bookmasters